Introduction to Cryptography

Sahadeo Padhye
Associate Professor
Department of Mathematics
Motilal Nehru National Institute of Technology
Allahabad, India

Rajeev A. Sahu
Post-Doctoral Researcher
Departement d'Informatique
Université Libre de Bruxelles (ULB)
Brussels, Belgium

Vishal Saraswat
Visiting Scientist
R.C. Bose Centre for Cryptology and Security
Indian Statistical Institute
Kolkata, India

CRC Press
Taylor & Francis Group
Boca Raton London New York

CRC Press is an imprint of the
Taylor & Francis Group, an **informa** business

A SCIENCE PUBLISHERS BOOK

CRC Press
Taylor & Francis Group
6000 Broken Sound Parkway NW, Suite 300
Boca Raton, FL 33487-2742

© 2018 by Taylor & Francis Group, LLC
CRC Press is an imprint of Taylor & Francis Group, an Informa business

No claim to original U.S. Government works

Printed on acid-free paper
Version Date: 20180524

International Standard Book Number-13: 978-1-138-07153-7 (Hardback)

Visit the Taylor & Francis Web site at
http://www.taylorandfrancis.com

and the CRC Press Web site at
http://www.crcpress.com

Dedication

To
our parents
and
our teachers

Foreword

Cryptography is an amazing science. It has been around for thousands of years, but took off phenomenal strides less than fifty years ago. At the crossroads of mathematics, computing, telecommunications networks and quantum physics, cryptography is nowadays a field of abundant researches and applications. Essential for the smooth running of the digital world in which we live, the learning of cryptography is therefore today a necessity for any IT specialist.

It is in this context that Sahadeo Padhye, Rajeev Anand Sahu and Vishal Saraswat offer us a new textbook on cryptography of very high quality. This book is an ideal companion for any student in computer science, in mathematics or in engineering who wishes to approach cryptography in a formal, scientific and rigorous manner. The authors have succeeded in proposing a textbook that presents cryptography in depth with a great deal of clarity.

This comprehensive and self-contained book introduces very clearly and iteratively all the mathematical elements necessary to understand the cryptographic tools presented afterwards. Indeed, the first five chapters will give readers the elements of algebra, number theory, probability and complexity of algorithms that will allow them to easily approach the next five chapters dedicated to the most fundamental tools of cryptography such as encryption, hash functions and digital signatures.

For every discussed cryptographic aspect, the reader will find informal explanations, formal definitions of all the underlying mathematical aspects, a clear description of the cryptographic scheme itself, numerical examples and corresponding exercises.

The reader will be taken on a journey from the oldest cryptographic algorithms to the most modern cryptographic schemes. From the scytale to elliptic curves cryptography.

Very complete, this book also presents really clear descriptions of most of the essential algorithms needed to implement cryptographic schemes, such as the computation of inverses, of square roots, primality tests, exponentiations, factorization, discrete logarithm, etc.

Few books deal with so many different aspects of general cryptography in such a didactic and comprehensive manner. All of this makes this introductory book on cryptography by Sahadeo Padhye, Rajeev Anand Sahu and Vishal Saraswat a teaching and pedagogical tool of quality, both for students or neophytes who are discovering this exciting science and for teachers who need to set up a new course on this subject.

Olivier Markowitch
Brussels, December 2017

Preface

In the past three decades, study and research in cryptography has been growing. 'Cryptography' is no longer an unknown term. In many universities students are studying cryptography in undergraduate and/or postgraduate courses, researchers are involved in exploring various useful primitives of cryptography and many people are using cryptography in their daily life, knowingly or unknowingly. Cryptography has been practiced since long, but it has attracted the attention of researchers and academicians in the last 30 years, since the elementary foundations proposed by IBM (the Data Encryption Standard) in mid 1970s, Diffie and Hellman (the key exchange protocol over the public network) in 1976 and by Rivest, Shamir and Adleman (the RSA public key cryptosystem) in 1978. The idea of cryptography is not independent of mathematics, in fact, major mathematical theories are at the core of cryptography. This book presents an introduction to the notion of cryptography and to the mathematical theories underlying the notion.

This book aims at readers who have some elementary background in mathematics and are interested to learn cryptography. We have introduced basic techniques of cryptography addressing the mathematical results used to establish those techniques. In this way we achieve one of our goals, i.e. to provide mathematical structures of basic cryptographic primitives and a list of ready reference to the students of undergraduate and postgraduate with a possibility to cover major sections of their syllabus of cryptography.

We have arranged the topics such that the mathematical theories have been described before being mentioned in design and/or analysis of cryptographic schemes. The book starts with an overview of cryptography which includes the goals of cryptography and cryptanalysis. In the second chapter, we have provided definitions from algebra that are essential in the formalization of various designs in later chapters. Number theory captures important results of mathematics used in cryptography. We have discussed many of these results with brief description in third chapter. Probability and random numbers are other important notions used in design and analysis of cryptographic algorithms. These theories have been covered in the fourth chapter. The fifth chapter introduces the theory of complexity, which plays important role in measure of strength of a cryptosystem. Thus, the first five chapters provide a mathematical foundation. Hereafter, we start explaining the concepts of cryptographic constructions. The explanation begins with description of the classical cryptosystems in chapter six. The description of modern cryptographic designs starts with details of symmetric key cryptography in chapter seven where we discuss various block ciphers. The asymmetric cryptosystems (public key cryptosystems) are explained in chapter nine. We have included description of digital signature, an important primitive of public key cryptography, in an independent chapter, chapter ten. Hash function is an integral ingredient of public key cryptography, it has been covered in chapters eight to ten. Chapter 11 discusses emerging topics in cryptography.

The authors thank everyone due to whom this book could have come in its current form. In particular, we are thankful to Vijay Primlani of Science Publishers CRC Press, for giving us the opportunity to publish this work. We express our special thanks to Prof. Olivier Markowitch who happily agreed to present his views about this book in the foreword. We are also grateful to Gaurav Sharma and Veronika Kuchta, postdoctoral researchers at ULB, Brussels, Belgium, for their useful suggestions during discussions. Sonika Singh and Swati Rawal, Ph.D. students of Sahadeo Padhye at MNNIT Allahabad, have extended their efforts in proofreading, we thank both of them. Last, but not least, we thank our respective wives and children for their patient and support during the writing of this book.

Sahadeo Padhye
Rajeev Anand Sahu
Vishal Saraswat

Contents

Chapter 1

Overview of Cryptography

CONTENTS

Cryptography is defined to be as an art and science of secret writing. In more intellectual discussions, the journey of cryptography started from *art of writing* and moved to the *science of writing*. Broadly speaking classical cryptography (the art) moved to modern cryptography (the digital science) in the late 20th century. In order to understand these views better, one needs to understand the notion of cryptography that dates back to the beginning of human civilization. There are extensive historical evidences which testify the existence of methods of secret communication in many ancient societies including Mesopotamia, Egypt, India, China and Japan, but details regarding the exact origin of cryptography is debatable. In the ancient period, cryptography was used through some meaningful symbols in a specific art form. Later, these symbols were converted in arranged logics in a scientific way for more sensitive and crucial purpose. The elementary cipher machines (e.g. Enigma Machine) used by military organizations, which played a vital role during world wars, were examples of the use of cryptography. After the precise algorithms and their implementations since 1980s, there was a paradigm shift in the era of cryptography and cryptography entered in more practical world of modern cryptography where its application influenced and involved everyone.

1.1 Introduction

The term cryptography is a combination of two Greek words "Kruptos" (i.e. hidden) and "Graphio" (i.e. writing). Formally, cryptography is the study of secure designs required for the protection of sensitive communication. A communication requires a channel for the transaction of messages. This channel may be a telephone network, internet connection, or something else. However, none of the existing channels are guaranteed to be secure. Cryptography is all about providing a secure communication over an insecure channel.

Cryptography deals with the means and methods of data conversation in a secure form so that the data cannot be used by unauthorized person, authenticity of the data can be established, modification on the data can be avoided and the data cannot be repudiated by the generator. A cryptographic protocol/algorithm used for the above purpose is called a *cryptosystem*. In cryptography, the original message is usually referred as *plaintext* and the text converted from plaintext into (usually) non-meaningful form is referred as *ciphertext*. In general, cryptography deals with encryption of a plaintext into ciphertext and decryption of ciphertext into plaintext. The keys for the conversion are called *encryption key* and *decryption key*. The study of ciphertext, in attempt to decrypt it or to recover the key, is called cryptanalysis. The term cryptography and cryptanalysis are not apart of each other and can be viewed as two sides of a single coin. In order to construct secure cryptography, one should design it against all possible cryptanalysis. The term *cryptology* refers to study of design and analysis of cryptosystems.

Below, we define a cryptosystem more formally as:

Definition 1.1 Cryptosystem: [162] A Cryptosystem is a five-tuple $(\mathcal{P}, \mathcal{C}, \mathcal{K}, \mathcal{E}, \mathcal{D})$, satisfying the following conditions-

1. $\mathcal{P}$ is a finite set of possible plaintexts;

2. $\mathcal{C}$ is a finite set of possible ciphertexts;

3. $\mathcal{K}$, the key space, is a finite set of possible keys;

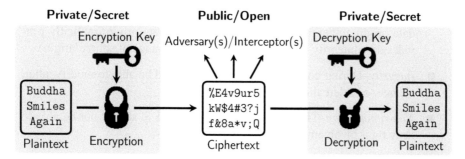

Figure 1.1: Cryptosystem

4. For each key $k \in \mathcal{K}$, there is an encryption rule $E_k \in \mathcal{E}$, $E_k : \mathcal{P} \rightarrow \mathcal{C}$ and a decryption rule $D_k \in \mathcal{D}$, $D_k : \mathcal{C} \rightarrow \mathcal{P}$ such that, for any plaintext $x \in \mathcal{P}$, $y = E_k(x)$ is the corresponding ciphertext and $D_k(E_k(x)) = x$.

It appears from the above definition that in a valid cryptosystem the encryption must be invertible since the decryption function is the inverse of the encryption function.

1.2 Goals of Cryptography

There are four basic goals of cryptography. These goals are abbreviated as 'PAIN', where P stands for **Privacy** (also known as confidentiality); A stands for **Authentication**; I stands for **Integrity**, and N stands for **Non-repudiation**. In addition to these basic goals, access control is now also considered to be one of the goals. We define each goal assuming a communication between a sender $\mathcal{S}$ and receiver $\mathcal{R}$.

■ *Privacy/Confidentiality*–No one except the authorized users $\mathcal{S}$ and $\mathcal{R}$ should be able to see/detect the message communicated between them. Secrecy is a term synonymous to privacy and confidentiality. In general we use the term privacy and confidentiality as synonyms. In particular, privacy refers to an individuals right to control access to their personal information where confidentiality refers to how private information provided by individuals will be protected from release.

■ *Authentication*–Authentication is a service related to identification. The receiver $\mathcal{R}$ should be able to validate the actual source or origin from where the message has been transmitted. Two parties involved in a communication should be able to identify each other. Also, the information delivered over the channel should be authenticated in the sense that it has been generated from an authorized source. Hence, authentication is required to be checked in two views: entity (or user) authentication and data origin authentication. Authentication of data origin implicitly provides data integrity (altered message implies change of source/origin).

■ *Integrity*–During communication no one should be able to modify, alter, change, or delete the message.

■ *Non-repudiation*–The actual participants S and R should not be able to deny their involvement in the communication.

■ *Access Control*–Only the legitimate users are granted access to the resources or information.

Different cryptographic primitives have been devised to achieve these goals. The privacy can be achieved by applying an encryption scheme. Entity authentication can be achieved by digital signature. Integrity can be achieved by hash function and the goal of non-repudiation can be achieved by advanced electronic signature or digital signature with some additional functionality (e.g. undeniable signature). We will precisely discuss each of these crypto-primitives in further chapters.

1.3 Classification of Cryptosystem

For a classified study, the cryptosystems can be categorized in the following classes:

■ *Classical vs Modern*–This classification is in the view of pre and post computer era. Classical cryptosystems are the ones which have been used during the pre-computer era and the modern cryptosystems are those being used in the post-computer era. Broadly, they can also be analyzed on the basis of time period i.e. those belonging to early twentieth century and those belonging to late twentieth century. Examples of classical cryptosystems are: Caesar Cipher, Vigenère Cipher, Affine Cipher, Hill Cipher, Auto-key cipher etc. and the examples of modern cryptosystems are DES, AES, RSA, ElGamal etc.

■ *Symmetric Cryptosystem vs Asymmetric Cryptosystem* or *Secret Key Cryptosystem (SKC) vs Public Key Cryptosystem (PKC)*–This classification is on the basis of encryption key e_k and decryption key d_k. If $e_k = d_k$ then the cryptosystem is called symmetric cryptosystem (or secret key crytosystem) and if $e_k \neq d_k$ then the cryptosystem is called asymmetric cryptosystem (or public key cryptosystem). Examples of symmetric cryptosystems are all classical cryptosystems, Block cipher, Stream cipher etc. Asymmetric cryptosystems are RSA, ElGamal, Knapsack, Rabin, XTR, NTRU etc.

■ *Block Cipher vs Stream Cipher*–This classification is on the basis of encryption process. In block cipher, the message is encrypted after being divided into fixed size of blocks. Examples of block ciphers are DES, AES, Triple DES, Serpent, Blowfish, Twofish, IDEA, RC5, Clefia etc. In stream cipher, the entire message is encrypted at a time using a key stream. RC4, ChaCha, SOSEMANUK, SEAL, SNOW, Grain, MICKEY, Trivium, Linear Feed Back Shift Registrar (LFSR) are some popular stream ciphers.

■ *Deterministic vs Randomized*–This classification is also based on the nature of encryption process. If the encryption formula always outputs the same ciphertext for a fixed plaintext then the cryptosystem is called deterministic cryptosystem. However, if the output of the encryption algorithm depends on some random number i.e. if encryption of the same message yields different ciphertext everytime depending upon the random number used in the encryption formula, then the cryptosystem is called randomized cryptosystem. Randomized cryptosystems are assumed to be more secure than deterministic cryptosystems. DES, AES, RSA, Rabin, XTR and others are deterministic cryptosystems and RSA-OAEP, ElGamal, NTRU etc. are randomized.

■ *Pre-quantum vs Post-quantum*–This classification is on the basis of the strength of the cryptosystem against quantum computer. The cryptosystems which are secure against quantum computer (quantum algorithm) are called post-quantum cryptosystems otherwise they are referred as pre-quantum cryptosystems. Examples of pre-quantum cryptosystems are all classical cryptosystems, RSA, ElGamal, Rabin etc. and the examples of post-quantum cryptosystems are Lattice-based cryptosystems (Ring-LWE, NTRU, GGH), Code-based cryptosystems (McEliece cryptosystem), Multivariate cryptography (Rainbow), Hash-based cryptography (Merkle signature scheme), Isogeny-based cryptosystem etc.

1.4 Practically Useful Cryptosystem

Following are two desired properties for a practically useful cryptosystem-

1. Each encryption function e_k and decryption function d_k should be easily computable.

2. An opponent, upon seeing a ciphertext string y, should be unable to determine the key k (that was used), or the plaintext string x.

The process of attempt to recover the key k or the message x given a string of ciphertext y is called *cryptanalysis*. If the attacker can determine the secret key, then the cryptosystem is totally broken and there is no security at all. Thus determining the key k is at least as difficult as determining the plaintext string x, given the ciphertext string y.

Sir Francis R. Bacon (1561–1626) had proposed the following properties for a "good" cryptosystem-

1. Given encryption key e_k and a plaintext x, it should be easy to compute ciphertext $y = e_k(x)$.

2. Given decryption key d_k and a ciphertext y, it should be easy to compute plaintext $x = d_k(y)$.

3. A ciphertext $e_k(x)$ should not be much longer than the plaintext x.

4. It should be infeasible to determine x from $e_k(x)$ without knowing d_k.

5. The *avalanche effect* should hold in the cryptosystem. Avalanche effect is an effect by which small changes in the plaintext or in the key lead to a big change in the ciphertext (i.e. a change of one bit in the plaintext usually results into changes in all bits of the ciphertext, each with the probability close to 0.5).

6. The cryptosystem should not be closed under composition, i.e. for any two keys k_1, k_2 there should not exist a key k such that $e_k(w) = e_{k_1}(e_{k_2}(w))$.

7. The set of keys should be very large.

1.4.1 Confusion and Diffusion

In 1949, Claude Shannon in one of his fundamental papers Communication Theory of Secrecy Systems explained two properties, confusion and diffusion, that a "good" cryptosystem should satisfy in order to hinder statistical analysis.

■ *Confusion*: Confusion implies that the key is not related in a simple or direct way to the ciphertext. If the confusion is high, then an attacker cannot extract information on plaintexts from the distribution of ciphertexts. The idea of confusion is to hide the relationship between the plaintext and the key. In particular, each character of ciphertext should depend on several parts of the key. Therefore, confusion is high if each bit of key has an influence on many bits of ciphertext.

■ *Diffusion*: When change on a character of the plaintext yields change in all characters of ciphertext and vice versa then it is called diffusion. The idea of diffusion is to hide the relationship between the ciphertext and the plaintext. So, the diffusion is high if each bit of a plaintext has an influence on many bits of ciphertext.

The concepts of confusion and diffusion play an important role in block cipher. A disadvantage of confusion and diffusion is error propagation which means a small error in the ciphertext reflects a major error in the decrypted ciphertext i.e. the original message. Hence, the message becomes unreadable.

1.5 Cryptanalysis

Cryptanalysis deals with the study, observations, and analysis of cryptosystem in order to attack on it. There are standard assumptions for cryptanalysis which relate its practicality in the real world. Dutch cryptographer Auguste Kerckhoffs has formalized standard guidelines to consider the attacker's privilege in design of a "good" cryptosystem. The assumption is popularly known as Kerckhoff's principle in cryptography. The principle says "A cryptosystem should be secure even if everything about the system, except the key, is publicly known". Thus, we test the security of a cryptosystem under the assumption that the security of the cryptosystem should depend on the secret key and not on the algorithm. In fact, the algorithm should be assumed to be publicly available.

1.5.1 Types of Attackers

When a sender Alice sends an encrypted message to Bob over a channel, then an attacker, Eve (eavesdropper), has the following privilege or access in the communication:

- Eve can read (and try to decrypt) the message.

- Eve can try to get the key that was used and then decrypt all encrypted messages with the same key. This is called total break.

- Eve can change the message sent by Alice into another message in such a way that Bob considers the message as if it was actually sent from Alice. This is called modification or attack on data integrity.

- Eve can pretend to be Alice and communicate with Bob in such a way that Bob thinks he is communicating with Alice. This is called masquerading or man-in-the-middle attack.

An eavesdropper/attacker can therefore be passive or active. A *passive attacker* can only read the message. Attacks that threaten confidentiality are passive attacks. *Active attacker* can even modify the message or can harm the system. Attacks that threaten the integrity and availability are active attacks.

Potential adversarial goals

- *Total break*–to determine the private/secret key.

- *Partial break*–to determine some specific information about plaintext or to decrypt a previously unseen ciphertext, with some non-negligible probability.

- *Distinguishability of ciphertexts*–to distinguish between encryption of two given plaintexts or between an encryption of a given plaintext and a random string with some probability exceeding $1/2$.

1.5.2 Types of Attacks

There are six types of attacks that Eve might be able to follow. The differences among these attacks are on the basis of the amounts of information that Eve can determine during his attempt to recover the secret key or the target plaintext. The six attacks are as follows-

1. Ciphertext only Attack–The cryptanalysts get ciphertexts $c_1 = e_k(m_1)$,..., $c_n = e_k(m_n)$ and try to infer the key k or the plaintexts $m_1, .., m_n$. i.e. Eve is given only a list of ciphertexts.

2. Known Plaintext Attack (KPA)–The cryptanalysts know some pairs $(m_i, e_k(m_i))$, $1 \leq i \leq n$ and try to infer k or at least m_{n+1} for a new ciphertext $c_{n+1} = e_k(m_{n+1})$. This is equivalent to the fact that the attacker can access the encryption oracle for messages known to him.

3. Chosen Plaintext Attack (CPA)–The cryptanalysts can choose plaintexts $m_1, .., m_n$ to get ciphertexts $e_k(m_1), ..., e_k(m_n)$ and try to infer k or at least m_{n+1} for a new ciphertext $c_{n+1} = e_k(m_{n+1})$. This is equivalent to the fact that the attacker can access the encryption oracle for messages chosen by him. In public key cryptosystem, adversary always has this privilege as in PKC encryption key is publicly available.

4. Chosen Ciphertext Attack (CCA)–The cryptanalysts can access some pairs $(c_i, d_k(c_i))$, $1 \leq i \leq n$, where the ciphertexts c_i have been chosen by the cryptanalysts and $d_k(c_i)$ are their decryption. This is equivalent to the fact that the attacker can access the decryption oracle for ciphertexts chosen by him.

5. Adaptive Chosen Plaintext Attack (CPA2)–The cryptanalysts can choose any plaintext m dynamically (i.e. the plaintext can be selected on the basis of response to the previous selection) and can get the corresponding ciphertext $c = e_k(m)$ except the target one. This is equivalent to the fact that the attacker can access the encryption oracle for messages chosen by him adaptively.

6. Adaptive Chosen Ciphertext Attack (CCA2)–The cryptanalysts can choose any ciphertext dynamically and can get the corresponding plaintext except the target one. This is equivalent to the fact that the attacker can access the decryption oracle for messages chosen by him adaptively.

In the symmetric cryptosystem the encryption and decryption keys are same. Hence, the key is secret. Without knowing the secret key no one can encrypt/decrypt any message. Therefore, in symmetric cryptosystem, attacker either attempts to find the plaintext for a target ciphertext (passive attack) or tries to find the ciphertext for a target plaintext (active attacker) to replace/modify the actual ciphertext. But in case of PKC, the encryption key is always publicly available. Hence, the attacker can encrypt any message and therefore there is no sense of CPA/CPA2 in public key encryption. Thus, we test the security of encryption in PKC against CCA or CCA2. But when we use PKC for authentication (digital signature) then message is the plaintext and signature on it is the ciphertext. Hence, analysis of digital signature against CPA/CPA2 makes sense.

1.5.3 Security Notions

How to measure/test the security of a cryptosystem? It is one of the most considerable issues in cryptography. Following are some of the various approaches to evaluate the security of a cryptosystem.

- **Computational Security**: Under this measure, we test the computational effort required to break a cryptosystem. A cryptosystem is computationally secure if the best known algorithm used by the current computer technology is unable to break the cryptosystem within a permissible time period. It is very difficult to define a permissible period, hence, no cryptosystem is defined to be secure under this definition. If the security of any cryptosystem is based on some hard (assumed) mathematical problem in number theory such as integer factoring problem, discrete logarithm problem in certain finite cyclic group, then such a cryptosystem is referred to be computationally secure as its solutions have not been achieved yet with the available resources.

- **Provable Security**: In this approach, we associate the security of a cryptosystem to some hard/intractable mathematical problem. We rely on the fact that the attack to the cryptosystem will be equivalent to solution of the hard mathematical problem which has not been solved yet in a permissible time frame.

■ **Unconditional Security**: A cryptosystem is defined to be unconditionally secure if it cannot be broken even with infinite computational resources. Under this measure, we provide infinite computational power to the attacker. This is categorically clear that an unconditional secure cryptosystem must be considered to be the strongest cryptosystem.

■ **Semantic Security**: In this approach, it is observed that whether the given ciphertext leaks any information about the plaintext or not. If the attacker cannot get any information about the plaintext from the target ciphertext then such a system is called semantically secure cryptosystem. All deterministic cryptosystems are not semantically secure. To achieve semantic security the system must be at least randomized. It should be noted that the randomized cryptosystems need not to be semantically secure. The semantic security is closely related to the ciphertext's distinguishability.

Chapter 2

Basic Algebra

CONTENTS

In this chapter, we will discuss some algebraic structures used in cryptography. Finite group, cyclic group, and finite field are some of the most important structures used in the design and analysis of cryptographic primitives. The well known RSA cryptosystem is defined over the group $Z_n^\star$, ElGamal cryptosystem is defined over the cyclic group $Z_p^\star$, and one of the popular symmetric key cryptosystem AES is defined over Galois field $GF(2^8)$.

Definition 2.1   Binary Operation: A binary operation $\star$ on a non-empty set S is a rule which associates any two elements of S into a third element of S.

Example 1 *Usual addition "+", subtraction "-" are binary operations on the set of integers Z, real numbers R etc.*

Remark 2.1 Any non-empty set S equipped with a binary operation $\star$ satisfying some conditions is called an algebraic structure. Some basic algebraic structures are Group, Ring and Field.

2.1 Group

Let G be any non-empty set and $\star$ be any binary operation defined on G. Then $(G, \star)$, an algebraic structure with binary operation $\star$, is called a group if it satisfies the following properties:

- **Closure:** $x \star y \in G$, where $x, y \in G$.
- **Associative:** $(x \star y) \star z = x \star (y \star z)$ where $x, y, z \in G$.
- **Existence of Identity Element:** There exists an element $e \in G$ such that $x \star e = e \star x = x$ for all $x \in G$. e is called identity element of G with respect to the operation $\star$.
- **Existence of Inverse Element:** There exists an element $y \in G$ such that $x \star y = e = y \star x$ for every $x \in G$. y is called inverse of x with respect to the operation $\star$.

If the structure $(G, \star)$ also satisfies the commutative property defined below

- **Commutative Property:** This means $x \star y = y \star x$ for all $x, y \in G$

then $(G, \star)$ is called an Abelian group after the name of the famous mathematician Niels Henrik Abel.

Remark 2.2 If $\star$ is a binary operation, then the closure property holds obviously.

Remark 2.3 Without the loss of generality we use the notation x^n for $\overbrace{x \star x \star x \star \ldots \star x}^{n\ times}$.

Example 2 *Following are some examples of group. All the properties can be easily verified for each of the examples.*

1. *The first example of group is $(R, +)$, the set of real numbers, with respect to usual addition.*

2. *The structure $(N, +)$, where $N = \{0, 1, 2, \ldots\}$, and $+$ is usual addition, does not form a group due to lack of existence of inverse of the elements.*

3. *The set of all 2×2 matrices forms a group under component-wise addition.*

4. *The structure $(Z_n^\star, \star)$ forms a group with respect to multiplication modulo n.*

5. $(R,+),(C,+),(Q,+),(R^n,+),(R^\star,\star),(C^\star,\star)$ *are examples of Abelian groups.*

6. *The set of permutations of n numers (denoted as S_n) is a group with respect to composition. For example, consider S_3, set of all permutations of numbers $(1,2,3)$. The composition operation is defined in the following table.*

∘	[1 2 3]	[1 3 2]	[2 1 3]	[2 3 1]	[3 1 2]	[3 2 1]
[1 2 3]	[1 2 3]	[1 3 2]	[2 1 3]	[2 3 1]	[3 1 2]	[3 2 1]
[1 3 2]	[1 3 2]	[1 2 3]	[2 3 1]	[2 1 3]	[3 2 1]	[3 1 2]
[2 1 3]	[2 1 3]	[3 1 2]	[1 2 3]	[3 2 1]	[1 3 2]	[2 3 1]
[2 3 1]	[2 3 1]	[3 2 1]	[1 3 2]	[3 1 2]	[1 2 3]	[2 1 3]
[3 1 2]	[3 1 2]	[2 1 3]	[3 2 1]	[1 2 3]	[2 3 1]	[1 3 2]
[3 2 1]	[3 2 1]	[2 3 1]	[3 1 2]	[1 3 2]	[2 1 3]	[1 2 3]

The shaded entry $[1,2,3]$ is the identity element of S_3 and the entries of corresponding rows and columns are inverse of each other.

Definition 2.2 **Order of a Group:** The number of distinct elements in a group is called the order of the group. It is also known as cardinality of the group. If the order is finite then the group is called finite group. We denote $\#(G) = n$ if the order of the group G is n.

Example 3 *Consider the group $(Z, +_n) = \{0, 1, 2, ..., n-1; +_n\}$, where $+_n$ denotes addition modulo n, a binary operation. The order of this group is n.*

Definition 2.3 **Cyclic Group:** A finite group G is called a cyclic group if G is generated by a single element $g \in G$. We denote this fact as $G = < g >$. The element g is called generator of G.

A finite cyclic group $G = < g >$ is necessarily an Abelian group. It can be expressed as $G = \{1, g, g^2, ..., g^{n-1}\}$, where $g^n = 1$ for the multiplicative group and as $G = \{0, g, 2g, ..., (n-1)g\}$, where $ng = 0$ for the additive group. Thus, a finite group with n elements is isomorphic to the additive group Z_n of integers modulo n.

Example 4 *See the following examples of cyclic group.*

1. *A group with prime order is always cyclic.*

2. *Z_n is cyclic group under addition modulo n.*

3. *The multiplicative group $Z_p^\star$ is cyclic group under multiplication modulo p, where p is a prime number.*

Remark 2.4 We denote a multiplicative group modulo n by $Z_n^\star$. For example $Z_{10}^\star = \{1, 3, 7, 9\}$.

Definition 2.4   Subgroup: A non-empty subset $H \subseteq G$ is called a subgroup of the group G if H itself forms a group under the same binary operation defined on G.

Example 5 $(Z, +)$ *is a subgroup of* $(R, +)$ *since* Z *is a subset of* R *and both are groups under usual addition.*

Example 6 $R^{\star}$ *is a subset of* R *but not a subgroup of* R *because both are groups under different operations.* $R^{\star}$ *is group under multiplication and* R *is group under addition.*

Example 7 *Let* G *be a group and* $a \in G$. *Then,* $H = < a >$ *is a subgroup of* G. *This* $H = < a >$ *is called cyclic subgroup of* G *with generator* a. *For example, let* $G = Z_{11}^{\star}$ *and* $H = < 3 > = \{3, 3^2 = 9, 3^3 = 5, 3^4 = 4, 3^5 = 1\}$ *is a cyclic subgroup of* $Z_{11}^{\star}$ *of order 5.*

Remark 2.5   A non-empty set $H \subseteq G$ is a subgroup of G if for all $x, y \in H$, $xy^{-1} \in H$.

Theorem 2.1
Lagrange's Theorem. *Assume that* G *is a finite group, and* H *is a subgroup of* G. *If the order of* G *and* H *are* $\#(G)$ *and* $\#(H)$ *respectively, then,* $\#(H)$ *divides* $\#(G)$.

Due to the space constraint we are omitting the proof of the theorem.

Definition 2.5   Order of an element: The *smallest* positive integer n, such that $x^n = e$, is called order of the element $x \in G$. Where e is the identity element in G. If no such positive integer n exists, then $x \in G$ is said to be an element of infinite order. We denote by $ord(x) = n$, if n is the order of x.

Example 8 *Consider the group* $(R^{\star}, \star)$. *In this group* $ord(1) = 1$ *and* $ord(-1) = 2$. *What will be the order of 7? 7 is an element of infinite order because there exists no positive* n *such that* $7^n = 1$.

Remark 2.6   In the group $(R^{\star}, \star)$, all elements other than 1 and -1 have infinite order.

Example 9 *Let us consider the group* $Z_7^{\star} = \{1, 2, 3, 4, 5, 6; \times_7\}$. *The binary operation defined on this group is multiplication modulo 7. As* $2 \in Z_7^{\star}$, *we see that* $2^1 = 1, 2^2 = 4, 2^3 = 8 \equiv 1 \mod 7$. *Thus the* $ord(2) = 3$.

Following are some useful results on cyclic group. We list them omitting their respective proofs.

Theorem 2.2
Every subgroup of a cyclic group is cyclic.

Theorem 2.3
For every positive divisor d of # < a >, < a > contains precisely one subgroup of order d.

Theorem 2.4
If # < a >= m, then # < a^k >= ord $(a^k) = \frac{m}{gcd(k,m)}$.

Theorem 2.5
For every positive divisor d of # < a >, < a > contains $\phi(d)$ elements of order d. Where $\phi(n)$ is count of total positive integers less than n and co-prime to n (e.g. $\phi(8) = 4$).

Theorem 2.6
Let # < a >= m. Then < a > contains $\phi(m)$ generators. They are elements a^r such that gcd $(r, m) = 1$.

Example 10 *Consider $Z_{11}^{\star}$. $\#Z_{11}^{\star} = 10$. We observe that $Z_{11}^{\star} = \{2, 2^2 = 4, 2^3 = 8, 2^4 = 5, 2^5 = 10, 2^6 = 9, 2^7 = 7, 2^8 = 3, 2^9 = 6, 2^{10} = 1\}$. Thus 2 is a generator of $Z_{11}^{\star}$. By the above theorem, total number of generators for $Z_{11}^{\star}$ are $\phi(10) = 4$. Other generators are given by 2^k; $gcd(k, 10) = 1$ i.e. $2^3 = 8, 2^7 = 7, 2^9 = 6$. The order of the subgroup $H =< 2^4 = 5 >$ is equal to $\frac{10}{2} = 5$.*

2.2   Ring

An algebraic structure $(R, +, \star)$ having two binary operations addition $(+)$ and multiplication $(\star)$ is called a Ring if it satisfies the following conditions:

1. R is an Abelian group with respect to addition $+$.

2. R satisfies closure property and associative property with respect to multiplication $\star$.

3. Distributive property of multiplication over addition holds i.e.

    (a) $x \star (y + z) = x \star y + x \star z$, $\forall x, y, z \in R$.
    (b) $(x + y) \star z = x \star z + y \star z$, $\forall x, y, z \in R$.

Remark 2.7   For the sake of simplicity, the multiplicative identity e is often denoted as 1 and the additive identity is denoted as 0.

Definition 2.6   Commutative Ring: If the ring R is commutative w.r.t. multiplication i.e. $x \star y = y \star x$ for all $x, y \in R$, then it is called commutative ring.

Definition 2.7 **Ring with unity:** If there exists an element $e(\neq 0) \in R$ such that $x \star e = x = x \star e$ for all $x \in R$, then ring is called ring with unity. The element e is called multiplicative identity.

Example 11 *Following are some examples of Ring:*

1. *The set Z of integers under ordinary addition and multiplication is a ring with unity.*

2. *The set Z_n under addition and multiplication modulo n is commutative ring with unity.*

3. *The set of all 2×2 matrices with integer entries is a non-commutative ring with unity.*

2.3 Field

An algebraic structure $(F, +, \star)$ is called a field if-

1. The set F is an Abelian group with respect to '$+$' operation.

2. The set F^0 is a commutative group with respect to '$\star$' operation, where $F^0 = F - \{0\}$.

3. Distributive property of multiplication over addition holds as follows:
$$x \star (y + z) = x \star y + x \star z$$
$$(x + y) \star z = x \star z + y \star z$$

Hence, a field is a non-zero commutative ring in which every non-zero element has its inverse.

Example 12 *Q, R, C are fields under usual addition and multiplication.*

Example 13 *Z_p is a field under addition and multiplication modulo p, where p is a prime number.*

Definition 2.8 **Without zero divisor:** A ring R is called without zero divisor if for any two elements $x, y \in R$ $x \star y = 0 \Rightarrow x = 0$ or $y = 0$.

Example 14 *Consider Z_{10}^{*}. Then $2 \neq 0$ and $5 \neq 0$ but $2 \times 5 = 10 \equiv 0 \mod 10$. Hence, Z_{10}^{*} is with zero divisor.*

Remark 2.8 A field is always without zero divisors.

Remark 2.9 Z_n, where n is a composite integer, is not a field since it contains zero divisors but for any prime p, Z_p is a field.

2.3.1 Finite Field

For each prime p and positive integer n there is, upto isomorphism, a unique field of order p^n. These fields are called Galois field named after its introducer Évariste Galois and are denoted as $GF(p^n)$.

Definition 2.9 **Subfield:** A subset E of a field F is called a subfield of F if E is itself a field with respect to the operations (addition and multiplication) defined in the field F. For example, $(R, +, .)$ is subfield of $(C, +, .)$.

Definition 2.10 **Prime field:** A field is called prime field if it has no proper subfield. For example, the rational field $(Q, +, .)$ is prime field, whereas the real field $(R, +, .)$ is not because Q is proper subset of R.

Definition 2.11 **Finite Field of Prime Order:** The simplest examples of finite fields are the fields of prime order. Let p be a prime number then Z_p, the integer modulo p, is the finite field of order p with addition and multiplication 'modulo p' as the field operations.

Remark 2.10 We denote the finite field of prime order by F_p.

Example 15 *The binary field $GF(2^n)$ is a finite field with 2^n elements. Each element of $GF(2^n)$ is a binary string of length n.*

Definition 2.12 **Homomorphism and Isomorphism:** Let (A, o) and $(B, \star)$ be two algebraic structures. A mapping $f : A \rightarrow B$ is homomorphism if $f(aob) = f(a) \star f(b)$, $\forall a, b \in A$. If the map f is also bijective (i.e. one-one and onto) then it is called isomorphism. If there exists an isomorphism mapping between two algebraic structures A and B then we say A and B are isomorphic and denote it by $A \cong B$.

Remark 2.11 If $f : A \rightarrow B$ is homomorphism and e is the identity element in A then $f(e)$ is the identity element in B, also $f(a^{-1}) = f(a)^{-1}$ for any $a \in A$. See the following examples.

Example 16 *$F_2 = \{0, 1, +, .\} \cong B = \{F, T, \wedge, \vee\}$. Where F_2 is the binary field and B is the logical field.*

Example 17 *For a prime number p, $(Z_{p-1}, +) \cong (Z_p^\star, .)$. The isomorphism mapping $f : Z_{p-1} \rightarrow Z_p^\star$ is given by $f(x) = g^x \mod p$, g is the generator of $Z_p^\star$.*

Definition 2.13 **Characteristic of a Field:** The least positive integer 'n' is called characteristic of a field F if $n.a = \overbrace{a + a + ... + a}^{n\,times} = 0$, where 0 is the additive identity of F.

Table 2.1: Irreducible Polynomials of degree upto 5

Degree	Irreducible Polynomials
1	$(x+1)$, (x)
2	(x^2+x+1)
3	(x^3+x^2+1), (x^3+x+1)
4	$(x^4+x^3+x^2+x+1)$, (x^4+x^3+1), (x^4+x+1)
5	(x^5+x^2+1), $(x^5+x^3+x^2+x+1)$, $(x^5+x^4+x^3+x+1)$, $(x^5+x^4+x^3+x^2+1)$, $(x^5+x^4+x^2+x+1)$

Remark 2.12   If no such n exists then F is said to have zero characteristic.

Remark 2.13   Every finite field has a prime characteristic.

2.3.2   Field Construction

In cryptography, many cryptosystems are defined over some finite field of a fixed order. The order of a finite prime field is equal to characteristic of the field. However, this is not the general case for every finite field. A more general form of finite field can be constructed using irreducible polynomials.

Definition 2.14   **Polynomial over an Algebraic Structure:** Let $(A, +, \star)$ be an algebraic structure. A polynomial over the set A is an expression of the form $f(x) = a_0 + a_1 x + a_2 x^2 + ... + a_{n-1}x^{n-1} + a_n x^n$, where $a_i \in A, n > 0$, and $a_n \neq 0$. Here n is called degree of the polynomial and a_n is called leading coefficient of the polynomial.

Remark 2.14   The set of all polynomials over A is denoted by $A[x]$.

Definition 2.15   **Irreducible Polynomial:** Let A be an algebraic structure, a polynomial $f \in A[x]$ is said to be irreducible over A if f has a positive degree and $f = g \times h$ with $g, h \in A[x]$ implies that either g or h is a constant polynomial.

Some Irreducible Polynomials: In the set of polynomials in $GF(2^n)$, the irreducible polynomial of degree n plays the role of modulus. Table 2.1 gives a set of irreducible polynomials of degree up to 5.

Remark 2.15   A polynomial is said to be reducible over A if it is not irreducible over A.

Example 18   *Consider the polynomial $x^2 + 1 = f(x) \in F[x]$. Clearly $x^2 + 1 = (x + i)(x - i)$, therefore, it is irreducible over Real field $(R, +, .)$ but reducible over Com-*

plex field $(C, +, .)$. *Thus, a polynomial may be reducible over some field but irreducible over the other field.*

2.3.3 Field Construction using Irreducible Polynomial

Let A be an algebraic structure and $f, g, q, r \in A[x]$ with $g \neq 0$, then we say r is the remainder of f when divided by g and is denoted by $r \equiv f \pmod{g}$. The set of the remainders of all polynomials in $A[x] modulo \ g$ is called the polynomials in $A[x]$ modulo g and is denoted by $A[x]_g$. Clearly, $A[x]_g$ is the set of all polynomials of degree less than $\deg(g)$.

Theorem 2.7
Let F be a field and f be a non-zero polynomial in $F[x]$, then $F[x]_f$ is a ring and is a field if and only if f is irreducible over F.

Theorem 2.8
Let F be a field with p elements and f be a 'degree-n' irreducible polynomial over F, then the number of elements in $F[x]_f$ is p^n.

Therefore, in order to construct a finite field of p^n elements, we consider an irreducible polynomial $f(x)$ of degree n over F_p. Then, $F[x]_f$ forms a field of p^n elements under the polynomial addition and multiplication modulo irreducible polynomial $f(x)$.

Theorem 2.9
For every prime p and for every integer $n > 0$, there exists a finite field of p^n elements.

Remark 2.16   For simplicity, we use the notation $GF(p^n)$ for $F_p[x]_f$ where f is an irreducible polynomial of degree n in F_p.

2.3.4 Galois Field GF(2^n)

Consider the binary field $F_2 = \{0, 1; \times_2, +_2\}$, where the binary operation is defined under addition and multiplication modulo 2. If $f(x)$ is an irreducible polynomial over F_2 of degree n, then, $F_2[x]_f$ forms a field with 2^n elements and is also denoted by $GF(2^n)$ or F_{2^n}.

Example 19 *We define a field GF(2^2) with 4 elements of 2-bit words. Addition $\oplus$ and Multiplication $\otimes$ are defined as:*

$\oplus$	00	01	10	11		$\otimes$	00	01	10	11
00	00	01	10	11		00	00	00	00	00
01	01	00	11	10		01	00	01	10	11
10	10	11	00	01		10	00	10	11	01
11	11	10	01	00		11	00	11	01	10

2.3.4.1 Integer Representation of Finite Field Elements

In this section, we discuss how to represent an element of a finite field (e.g., F_{2^8}) as an integer in binary string. Consider an irreducible polynomial $f(x) = x^8 + x^4 + x^3 + x + 1$ over F_2. The set of all polynomials modulo $f(x)$ over F_2 forms a field of 2^8 elements i.e. $F_2[x]_f = \{g(x) = b_7x^7 + b_6x^6 + b_5x^5 + b_4x^4 + b_3x^3 + b_2x^2 + b_1x + b_0; b_i \in F_2\}$. Thus any element in this field can be represented as an integer of 8 binary bits $b_7b_6b_5b_4b_3b_2b_1b_0$ or a byte and vice versa.

In hexadecimal representation, we denote an integer by 4 bits as $0 = 0000(= 0), 1 = 0001(= 1), ...9 = 1001(= 9), A = 1010(= 10), ..., F = 1111(= 15)$. The Hexadecimal representation of a byte can use two quoted characters 'XY' such that $0 \leq X, Y \leq F$, i.e. any element in the field $F_2[x]_f$ can be viewed as a byte in the interval ['00', 'FF'].

Example 20 *Suppose we have 8-bit word 10011001. Its polynomial representation is $1.x^7 + 0.x^6 + 0.x^5 + 1.x^4 + 1.x^3 + 0.x^2 + 0.x + 1.x^0 = x^7 + x^4 + x^3 + 1$. Thus, $10011001 = x^7 + x^4 + x^3 + 1$.*

Polynomial Addition, Multiplication and Division in $GF(2^n)$: For the addition of two polynomials in $GF(2^n)$, we use $\oplus$ (XOR) operation componentwise. x^{i+j} is the multiplication of x^i and x^j. In multiplication, when the exponent exceeds $n-1$, then it is reduced under modulo irreducible polynomial by applying modulo 2 operation in the coefficients.

Example 21 *Evaluate $(x^7 + x^3 + x) \oplus (x^6 + x^3 + 1)$ in $GF(2^8)$. The symbol $\oplus$ represents the polynomial addition in $GF(2^8)$. The procedure is as follows:*

	$1x^7 + 0x^6 + 0x^5 + 0x^4 + 1x^3 + 0x^2 + 1x + 0$	
$\oplus$	$0x^7 + 1x^6 + 0x^5 + 0x^4 + 1x^3 + 0x^2 + 0x + 1$	
	$1x^7 + 1x^6 + 0x^5 + 0x^4 + 0x^3 + 0x^2 + 1x + 1$	$\rightarrow x^7 + x^6 + x + 1$

Example 22 *Evaluate $f(x) \otimes g(x)$ where $f(x) = (x^7 + x^3 + x)$ and $g(x) = (x^6 + x^3 + 1)$ in $GF(2^8)$. The symbol $\otimes$ represents the polynomial multiplication in $GF(2^8)$.*

The procedure is as follows

$$
\begin{aligned}
f(x) \otimes g(x) &= (x^7 + x^3 + x) \otimes (x^6 + x^3 + 1) \\
&= x^7(x^6 + x^3 + 1) + x^3(x^6 + x^3 + 1) + x(x^6 + x^3 + 1) \\
&= x^{13} + x^{10} + x^7 + x^9 + x^6 + x^3 + x^7 + x^4 + x \\
&= x^{13} + x^{10} + x^9 + x^6 + x^4 + + x^3 + x \\
&= x^{13} + x^{10} + x^9 + x^6 + x^4 + + x^3 + x \quad \bmod x^8 + x^4 + x^3 + x + 1 \\
&= x^6 + x^3 + x^2 + 1
\end{aligned}
$$

The division of $x^{13} + x^{10} + x^9 + x^6 + x^4 + + x^3 + x$ by $x^8 + x^4 + x^3 + x + 1$ is done as usual. We can write $x^{13} + x^{10} + x^9 + x^6 + x^4 + + x^3 + x = (x^5 + x^2 + 1) \times (x^8 + x^4 + x^3 + x + 1) + x^6 + x^3 + x^2 + 1$.

Inverse of Polynomial with coefficients in F_2: The well known cryptosystem AES is defined over the Galois Field $GF(2^8)$. During encryption and decryption we need to compute inverse of a byte. To find inverse of the byte, we first find the polynomial representation of the byte and then find the inverse of the polynomial under the given irreducible polynomial. Extended Euclidean algorithm (see Chapter 3, Algorithm 2) can be used to find inverse of a polynomial in $GF(2^n)$. Let us see the following examples.

Example 23 *The inverse of $x^2 + 1$ modulo $x^4 + x + 1$ in $GF(2^4)$ is $x^3 + x + 1$ as follows*

i	-1	0	1	2	3
r_i	$x^4 + x + 1$	$x^2 + 1$	x	1	0
q_i			$x^2 + 1$	x	x
μ_i	0	1	$x^2 + 1$	$x^3 + x + 1$	

Example 24 *Find the inverse of 00100000 in $GF(2^8)$, where the irreducible polynomial to define $GF(2^8)$ is $x^8 + x^4 + x^3 + x + 1$. The polynomial representation of 00100000 is x^5. Now we find the inverse of x^5 under modulo $x^8 + x^4 + x^3 + x + 1$ in $GF(2^8)$ using Extended Euclidian algorithm as below.*

i	-1	0	1	2	3	4
r_i	$x^8 + x^4 + x^3 + x + 1$	x^5	$x^4 + x^3 + x + 1$	$x^3 + x^2 + 1$	1	0
q_i			x^3	$x + 1$	x	$x^3 + x^2 + 1$
μ_i	0		1	x^3	$x^4 + x^3 + 1$	$x^5 + x^4 + x^3 + x$

Thus the inverse of 00100000 is 00111010. i.e. $x^5 + x^4 + x^3 + x$ in $GF(2^8)$.

2.3.5 Field Construction using Generator

We have seen that under isomorphism, $F_p[x]_f$ is the finite field of order $p^{deg(f)}$. However, often it may not be convenient to use fields modulo an irreducible polynomial. Multiplication in the field $F_p[x]_f$ is a bit complicated and needs modulo polynomial which requires the Euclid algorithm for polynomial division. In this section, we construct finite fields using the roots of an irreducible polynomial over a finite field F. Fields constructed in this way are frequently used in many applications as the computation becomes easier. If $f(x)$ is an irreducible polynomial over F_p of degree n and g is a root of $f(x)$ so that $f(g) = 0$, then the elements of $F_p[x]_f$ can be expressed as $\sum_{i=0}^{n-1} b_i g^i$, where $b_i \in F_p$. We denote this field by $GF(p^n)$ or F_{p^n}. In particular, consider the field F_{2^8} as explained.

Field F_{2^8}: Consider the irreducible polynomial $f(x) = x^8 + x^4 + x^3 + x + 1$ of degree 8 over F_2. Then $F_2[x]_f$ is the set of all polynomials modulo the irreducible polynomial $f(x)$ over F_2 and has 2^8 elements. These 2^8 elements can be represented as $\{b_7 g^7 + b_6 g^6 + b_5 g^5 + b_4 g^4 + b_3 g^3 + b_2 g^2 + b_1 g + b_0 : b_i \in F_2\}$, where g is a root of the equation $x^8 + x^4 + x^3 + x + 1 = 0$. We denote this field by F_{2^8}. Clearly, these two fields $F_2[x]_f$ and F_{2^8} are isomorphic. Multiplication in the field F_{2^8} is easier. Let us see the following example.

Example 25 *Let us compute* $(g^6 + g^4 + g^2 + g + 1).(g^7 + g + 1) = g^{13} + g^{11} + g^9 + g^8 + g^6 + g^5 + g^4 + g^3 + 1$. *Since* $g^8 + g^4 + g^3 + g + 1 = 0$. *We have the following linear combination(using 1 =-1 in F_2).*

$$
\begin{aligned}
g^8 &= g^4 + g^3 + g + 1 \\
g^9 &= g^5 + g^4 + g^2 + g \\
g^{11} &= g^7 + g^6 + g^4 + g^3 \\
g^{13} &= g^9 + g^8 + g^6 + g^5 + g^3
\end{aligned}
$$

Thus, $g^{13} + g^{11} + g^9 + g^8 + g^6 + g^5 + g^4 + g^3 + 1 = g^7 + g^6 + 1$.

We can also define the elements of the field $GF(2^n)$ by using the root g as a generator of the field as follows. $GF(2^n) = \{0, g^0, g^1, g^2, ..., g^N\}$, where $N = 2^n - 2$. Let us see the following examples.

Example 26 *Generate the elements of the field $GF(2^4)$ by using the irreducible polynomial $x^4 + x + 1$ as a generator.*

When g be a root of $x^4 + x + 1$ i.e. $g^4 + g + 1 = 0 \Rightarrow g^4 = g + 1$, then
$0 \longrightarrow 0 = (0000)$
$g^0 = 1 \longrightarrow g^0 = (0001)$
$g^1 = g \longrightarrow g^1 = (0010)$
$g^2 = g^2 \longrightarrow g^2 = (0100)$
$g^3 = g^3 \longrightarrow g^3 = (1000)$

$g^4 = g^4 = g + 1 \longrightarrow g^4 = (0011)$
$g^5 = g(g^4) = g(g+1) = g^2 + g \longrightarrow g^5 = (0110)$
$g^6 = g(g^5) = g(g^2 + g) = g^3 + g^2 \longrightarrow g^6 = (1100)$
$g^7 = g(g^6) = g(g^3 + g^2) = g^3 + g + 1 \longrightarrow g^7 = (1011)$
$g^8 = g(g^7) = g(g^3 + g + 1) = g^2 + 1 \longrightarrow g^8 = (0101)$
$g^9 = g(g^8) = g(g^2 + 1) = g^3 + g \longrightarrow g^9 = (1010)$
$g^{10} = g(g^9) = g(g^3 + g) = g^2 + g + 1 \longrightarrow g^{10} = (0111)$
$g^{11} = g(g^{10}) = g(g^2 + g + 1) = g^3 + g^2 + g \longrightarrow g^{11} = (1110)$
$g^{12} = g(g^{11}) = g(g^3 + g^2 + g) = g^3 + g^2 + g + 1 \longrightarrow g^{12} = (1111)$
$g^{13} = g(g^{12}) = g(g^3 + g^2 + g + 1) = g^3 + g^2 + 1 \longrightarrow g^{13} = (1101)$
$g^{14} = g(g^{13}) = g(g^3 + g^2 + 1) = g^3 + 1 \longrightarrow g^{14} = (1001)$

Clearly, the field $GF(2^4)$ *can be expressed as* $GF(2^4) = \{\sum_{i=0}^{i=3} b_i g^i : b_i \in (0,1)\}$. *Here, each* $b_3 g^3 + b_2 g^2 + b_1 g + b_0$ *corresponds to some* g^i, $0 \leq i \leq 14$.

We can use finite field $GF(2^n)$ to define four operations addition, subtraction, multiplication and division over n bit word.

Remark 2.17 The order of the multiplicative group $GF(2^n)^\star$ is $2^n - 1$. Hence, for any $k > 2^n - 1$, $g^k = g^{k \mod (2^n - 1)}$.

Example 27 *Using the above representation, the four basic operations can be defined as below.*

1. *Addition -* $g^3 + g^{12} + g^7 = g^3 + (g^3 + g^2 + g + 1) + (g^3 + g + 1) = g^3 + g^2 \longrightarrow$ (1100).

2. *Subtraction -* $g^3 - g^6 = g^3 + g^6 = g^3 + (g^3 + g^2) = g^2 \longrightarrow$ (0100).

3. *Multiplication -* $g^9 \times g^{12} = g^{20} = g^{20 \mod 15} = g^5 = g^2 + g \longrightarrow$ (0110).

4. *Division -* $\frac{g^3}{g^8} = g^3 \times g^{-8} = g^{3+7} = g^{10} = g^2 + g + 1 \longrightarrow$ (0111).

2.4 Exercises

1. Compute the order of the following groups and order of their elements
   (a) $\mathbb{Z}_7^*$ (b) $\mathbb{Z}_{11}^*$.

2. Prove that group $\mathbb{Z}_{15}^*$ is not cyclic under multiplication mod 15 .

3. Represent the following 8-bit words as polynomials in $GF(2^8)$
   (a) 00100110 (b) 00101011.

4. List all the irreducible polynomials in $GF(2^4)$ and $GF(2^5)$.

5. Add the following polynomials in $GF(2^8)$
   (a) $x^5 + x^2 + 1$ and $x^7 + x^6 + x^3 + x^2 + 1$
   (b) $x^4 + x^3 + x^2 + x + 1$ and $x^5 + x^4 + x^2 + x + 1$.

6. Show that the addition of polynomials in question 6 is same as XOR operation of their corresponding 8-bit representations.

7. Compute the following in $GF(2^8)$ with irreducible polynomial $x^8 + x^4 + x^3 + x + 1$
   (a) $(x^5 + x^3 + x^2 + x + 1) \times (x^4 + x + 1)$
   (b) $x^12 + x^7 + x^2 + x + 1 \mod x^8 + x^4 + x^3 + x + 1$

8. Find the inverse using Extended Euclidean Algorithm of following
   (a) $x^3 + x + 1 \mod x^4 + x + 1$
   (b) $x^7 \mod x^8 + x^4 + x^3 + x + 1$
   (c) $x^5 + x^4 + x^3 + x \mod x^8 + x^4 + x^3 + x + 1$
   (d) $x^4 + x^3 + 1 \mod x^5 + x^2 + 1$.

9. Find the primitive root in $\mathbb{Z}_7^*$ and show that there are no primitive roots in $\mathbb{Z}_8^*$ under multiplication modulo.

10. Prove that the multiplicative group Z_n^* is not cyclic, where $n = pq$ is a product of two distinct odd primes.

Chapter 3

Number Theory

CONTENTS

Number theory and Cryptography are inextricably linked. Most of the public key cryptosystems find their background in number theory as we will see in the further sections of this chapter. In this chapter, we will discuss some basic concepts of number theory in the context of cryptography. We will majorly focus on the discussion of primality testing, factoring problem, quadratic residuosity, and discrete logarithm problem.

3.1 Prime Numbers

Definition 3.1 **Prime Number:** A positive integer is called prime if and only if it is exactly divisible by two integers, 1 and itself.

The smallest prime is 2. The first few primes are $2, 3, 5, 7, 11, 13, 17, 19, \ldots$. As of May 2017, the largest known prime number is $2^{74,207,281} - 1$, which is a number of 22,338,618 digits. It was found in 2016 by the Great Internet Mersenne Prime Search (GIMPS).

Definition 3.2 **Composite Number:** A number n is composite if it is not prime. First few composite numbers are 4,6,8,9,10,12,14,15,.. .

Remark 3.1 The number 1 is neither prime nor composite.

Definition 3.3 **Relatively Prime or Co-prime:** Two positive integers a and b are said to be relatively prime if $\gcd(a, b) = 1$, where gcd means greatest common divisor.

Definition 3.4 **Greatest Common Divisor (GCD):** If $a = p_1^{a_1} \times p_2^{a_2} \times \ldots p_k^{a_k}$ and $b = p_1^{b_1} \times p_2^{b_2} \times \ldots p_k^{b_k}$, then the greatest common divisor of a and b is given by $\gcd(a,b) = p_1^{min(a_1,b_1)} \times p_2^{min(a_2,b_2)} \times \ldots p_k^{min(a_k,b_k)}$.

Definition 3.5   Least Common Multiple (LCM): If $a = p_1^{a_1} \times p_2^{a_2} \times ... p_k^{a_k}$ and $b = p_1^{b_1} \times p_2^{b_2} \times ... p_k^{b_k}$, then the least common multiple of a and b is given by $lcm(a,b) = p_1^{max(a_1,b_1)} \times p_2^{max(a_2,b_2)} \times ... p_k^{max(a_k,b_k)}$.

Relation between GCD and LCM of two numbers: The relation between GCD and LCM of two numbers is $gcd(a,b) \times lcm(a,b) = a \times b$.

3.2   Cardinality of Primes

People are generally curious to know that how many prime numbers actually exist. The answer is "infinitely many". An informal proof of this statement is as follows. Let us assume, contrary to the aforementioned statement, that the number of primes is finite, say p. Now multiply the set of all primes less than and equal to p. We get the result $P = 2 \times 3 \times 5 \times ... \times p$, clearly no primes $q \leq p$ will divide $P + 1$. As we know that if any prime $q \leq p$ divides $P + 1$, then that q must divide P also. Hence, $q|(P + 1) - P = 1$. The number which divides 1 is 1 itself, which is not a prime number. Hence, either $P + 1$ itself is a prime or having a prime factor not belonging to P. Consequently there exist infinite prime numbers. The next question is how many primes exist which are less than a given number n?

Theorem 3.1
Prime Number Theorem: Suppose $\pi(n)$ be the number of primes less than n, then $\pi(n) \sim \frac{n}{lnn}$, where $\sim$ denotes that the ratio $\frac{\pi(n)}{\frac{n}{lnn}} \to 1$ as $n \to \infty$.

We can estimate the number of 100-digit primes as follows:

$$\pi(10^{100}) - \pi(10^{99}) \sim \frac{10^{100}}{ln10^{100}} - \frac{10^{99}}{ln10^{99}} \sim 3.9 \times 10^{97}$$

Uptil now, we discussed that there are many prime numbers which exist under the given upper bound. In cryptography, many times we need to generate prime number of given size. In 1972, Euler first noticed that the quadratic polynomial $P(n) = n^2 + n + 41$ is prime for $n = 0, 1, 2, ..., 39$. The primes for $n = 0, 1, 2, ..., 39$ are $41, 43, 47, 53, 61, 71, ..., 1601$. For $n = 40$, it produces a square number, $1681 = 41 \times 41$. Some other mathematicians (Mersenne and Fermat) also tried to generate some special type of primes satisfying a given formula. They are called Mersenne primes and Fermat's primes.

Definition 3.6   Mersenne Primes: A number of the form $M_p = 2^p - 1$ is called Mersenne number. If p is prime, then it was observed that M_p be a prime number. Such primes are called Mersenne prime after the name of French mathematician Marin Mersenne. Later, it was proved that not all M_p are primes.

Some first Mersenne primes are $M_2 = 2^2 - 1 = 4 - 1 = 3, M_3 = 2^3 - 1 = 8 - 1 = 7$, $M_5 = 2^5 - 1 = 32 - 1 = 31, M_7 = 2^7 - 1 = 128 - 1 = 127, M_{11} = 2^{11} - 1 = 2048 - 1 = 2047$ (M_{11} is not a prime since $2047 = 23 \times 89$, hence, this is the smallest Mersenne composite number), $M_{13} = 2^{13} - 1 = 8192 - 1 = 8191$ etc. As of January 2016, 49 Mersenne primes are known. The largest known prime number $2^{74,207,281} - 1$ is a Mersenne prime.

Definition 3.7   Fermat Primes: The prime numbers of the form $F_n = 2^{2^n} + 1$ (i.e. Fermat number) are called Fermat primes. As they are just 1 more to a square number, they are also known as "near-square primes".

The only known Fermat primes are first five Fermat numbers. These Fermat primes are - $F_0 = 2^{2^0} + 1 = 3, F_1 = 2^{2^1} + 1 = 5, F_2 = 2^{2^2} + 1 = 17, F_3 = 2^{2^3} + 1 = 257, F_4 = 2^{2^4} + 1 = 65537, F_5 = 2^{2^5} + 1 = 4294967297$ (not prime since $4294967297 = 641 \times 6700417$).

Definition 3.8   Mersenne-Fermat Primes: A Mersenne-Fermat number is defined as $\dfrac{2^{p^r} - 1}{2^{p^{r-1}} - 1}$, where p is a prime number and r is a natural number. The Mersenne-Fermat prime is denoted as $MF(p, r)$. For $r = 1$, it is a Mersenne number, and for $p = 2$, it is a Fermat number. The only known Mersenne-Fermat prime with $r > 1$ are $MF(2, 2), MF(3, 2), MF(7, 2), MF(59, 2), MF(2, 3), MF(3, 3), MF(2, 4)$, and $MF(2, 5)$.

Definition 3.9   Euler's phi-Function $\phi(n)$: This function $\phi(n)$ is the number of integers that are smaller than n and relatively prime to n. It is also named as Euler's totient-function. The Euler's phi-function plays an important role in cryptography, particularly in the factoring based public key cryptosystem (PKC) and DLP based PKC like RSA and ElGamal.

Following are some properties of $\phi(n)$:

1. $\phi(1) = 1$.

2. If n is prime, then $\phi(n) = n - 1$.

3. $\phi(n)$ has multiplicative property i.e. $\phi(m \times n) = \phi(m) \times \phi(n)$ where m and n are relatively prime.

4. If n is prime, then $\phi(n^e) = n^e - n^{e-1}$.

The fundamental rule of arithmetic says that a positive integer n can be expressed as $n = p_1^{e_1} \times p_2^{e_2} \times ... p_k^{e_k}$ where $p_1, p_2, ..., p_k$ are the prime divisors of n. Then, $\phi(n) = n \times (1 - \frac{1}{p_1}) \times (1 - \frac{1}{p_1}) \times ...(1 - \frac{1}{p_k}) = (p_1^{e_1} - p_1^{e_1-1})(p_2^{e_2} - p_2^{e_2-1})...(p_k^{e_k} - p_k^{e_k-1}) = \phi(p_1^{e_1})\phi(p_2^{e_2})...\phi(p_k^{e_k})$.

If $n > 2$, then $\phi(n)$ is always even.

Example 28   *Using the property of $\phi(n)$, we can calculate the value of $\phi(n)$ as given in the following examples.*

1. $\phi(11) = 11 - 1 = 10$.

2. $\phi(10) = \phi(2) \times \phi(5) = 1 \times 4 = 4$.

3. $\phi(180) = \phi(3^2) \times \phi(2^2) \times \phi(5) = (3^2 - 3) \times (2^2 - 2) \times (5 - 1) = 6 \times 2 \times 4 = 48$.

4. *The number of elements in $Z_{18}^{\star}$ are* $\phi(18) = \phi(2) \times \phi(3^2) = 1 \times 6 = 6$.

Theorem 3.2
For a positive integer $n > 0$, $\sum_{d|n} \phi(d) = n$.

Proof 3.2   Let $S = (1, 2, 3, ..., n)$ and $S_d = (x | 1 \leq x \leq n, gcd(x, n) = d)$, where S_d contains all multiple of d less than n. Clearly , $\cup_{d|n} S_d = S \Rightarrow \sum_{d|n} \phi(\frac{n}{d}) = n \Rightarrow \sum_{\frac{n}{d}|n} \phi(\frac{n}{d}) = n$ (as for any $d|n, \frac{n}{d}|n) \Rightarrow \sum_{d|n} \phi(d) = n$.

Example 29   *For $n = 18$, all possible $d = 1, 2, 3, 6, 9, 18$. Then, $S_1 = \{1, 5, 7, 11, 13, 17\}$, $S_2 = \{2, 4, 8, 10, 14, 16\}, S_3 = \{3, 12, 15\}, S_6 = \{6\}, S_9 = \{9\}, S_{18} = \{18\} \Rightarrow \phi(1) + \phi(2) + \phi(3) + \phi(4) + \phi(6) + \phi(9) + \phi(18) = 6 + 6 + 3 + 1 + 1 + 1 = 18$.*

Definition 3.10   **Primitive Root:** For $m > 1$, an integer g is called a primitive root modulo m if $g^{\phi(m)} \equiv 1 \mod m$.

3.3   Extended Euclidean Algorithm

Euclidean Algorithm: This algorithm finds gcd of two numbers. If a and b are two positive integers with $a > b$, then $gcd(a, b) = gcd(b, a \mod b)$.

Algorithm 1 Euclidean Algorithm

INPUT: Two non-negative integers a and b with $a > b$.

OUTPUT: $\gcd(a,b)$.

1. **While** $b \neq 0$ do the following:

1.1 Set $r \leftarrow a \mod b$, $a \leftarrow b$, $b \leftarrow r$;

2. Return (a).

Time Complexity: The time complexity of Algorithm 1 is $\mathcal{O}(log\ a\ log\ b)$.

Example 30 *The* $\gcd(2261,1275)$ *can be computed using Euclidean Algorithm as following:*

$2261 = 1 \times 1275 + 986,\ 1275 = 1 \times 986 + 289$

$986 = 3 \times 289 + 119,\ 289 = 2 \times 119 + 51$

$119 = 2 \times 51 + 17,\ 51 = 3 \times 17 + 0$

We can observe that $\gcd(2261,1275) = \gcd(1275,986) = \gcd(986,289) = \gcd(289,119) = \gcd(119,51) = \gcd(51,17) = 17$. *Hence,* $\gcd(2261,1275) = 17$.

Extended Euclidean Algorithm: Given two integers a and b, there exist two integers λ and μ such that $\lambda \times a + \mu \times b = \gcd(a,b)$.

Algorithm 2 Extended Euclidean Algorithm

INPUT- Two non-negative integers a and b with $a > b$;

OUTPUT- Integers λ and μ satisfying $a\lambda + b\mu = \gcd(a,b)$;

1. $i \leftarrow 0; r_{-1} \leftarrow a; r_0 \leftarrow b$;

$\lambda_{-1} \leftarrow 1; \lambda_0 \leftarrow 0; \mu_{-1} = 0; \mu_0 \leftarrow 1$;

2. **while** $(r_i = a\lambda_i + b\mu_i \neq 0)$

do

(a) $q \leftarrow r_{i-1}\%r_i$;

(b) $\lambda_{i+1} \leftarrow \lambda_{i-1} - q\lambda_i$; $\mu_{i+1} \leftarrow \mu_{i-1} - q\mu_i$;

(c) $i \leftarrow i+1$;

3. **Return** $(\lambda_{i+1}, \mu_{i+1})$.

Time Complexity: It is clear that the number of recursive calls in Euclidean Algorithm is equal to the number of loops in Extended Euclidian Algorithm. Hence, time complexity of both the algorithms is same i.e. $\mathcal{O}(log\ a\ log\ b)$.

The extended Euclidean Algorithm can calculate the gcd (a,b) and also the values of λ and μ in the time required to calculate gcd (a,b).

Example 31 *Consider $a = 15, b = 12$, then gcd $(12,15) = 3$ and $1 \times 15 + (-1) \times 12 = 3$. Here, $\lambda = 1$ and $\mu = -1$ can be computed using the Algorithm 2 as below.*

$i \rightarrow$	-1	0	1	2
r_i	15	12	3	0
q_i			1	4
λ_i	1	0	1	-4
μ_i	0	1	-1	5

If $gcd(a,b) = 1$ where $a > b$, then by the relation $\lambda a + \mu b = gcd(a,b) \Rightarrow \mu b = 1$ mod a. Thus, $b^{-1} = \mu$ mod a. In this way, we can compute the *inverses* of a number under b mod a if $gcd(b,a) = 1$. This is one of the important applications of the extended Euclidean Algorithm in cryptography.

Example 32 *Let us find the inverse of 15 mod 26. Here $a = 26$ and $b = 15$. Since gcd $(15, 26) = 1$, inverse of 15 exists. Using the Extended Euclidean Algorithm, we find $4 \times 26 + 7 \times 15 = 1$. Thus, 15^{-1} mod $26 = 7$.*

Fermat's Little Theorem: If p is a prime and a is an integer such that p does not divide a, then $a^{p-1} \equiv 1$ mod p or we can say that $a^p \equiv a$ mod p.

Fermat's Little Theorem can be applied to find multiplicative inverse in prime modulus and in probabilistic primality testing.

Example 33 *Let $a = 2$ and $p = 7$, then, $a^{p-1} = 2^{7-1} = 64 \equiv 1 \pmod 7$.*

To compute the exponentiation a^k mod p, where k is a large number, we can write $a^k = a^{(p-1)q+r}$ mod $p \equiv a^r$ mod p, where $r < p$. Let us see the following example.

Example 34 *Find the value of 3^{22} mod 11. We can write this as $3^{2 \times 10 + 2}$ mod 11. Now using Fermat's Little Theorem, it will be equal to 3^2 mod $11 = 9$ mod 11. Thus, 3^{22} mod $11 = 9$.*

Multiplicative inverse using Fermat's Little Theorem: We can find multiplicative inverses modulo a prime without using extended Euclidean Algorithm. By Fermat's theorem, we have a^{-1} mod $p \equiv a^{p-1}a^{-1} \equiv a^{p-2}$ mod p.

Example 35 *Now, we shall find the inverse of some numbers using Fermat's theorem.*

1. 17^{-1} mod $61 \equiv 17^{61-2}$ mod $61 \equiv 17^{51}$ mod $61 \equiv 18$ mod 61.

2. 37^{-1} mod $101 \equiv 37^{101-2}$ mod $101 \equiv 37^{99}$ mod $101 \equiv 71$ mod 101.

3. 22^{-1} mod $211 \equiv 22^{211-2}$ mod $211 \equiv 22^{209}$ mod $211 \equiv 48$ mod 211.

Euler's Theorem: If a and n are relatively prime, then $a^{\phi(n)} \equiv 1 \pmod{n}$ or if $a < n$ and k is an integer, then $a^{k \times \phi(n)+1} \equiv a \mod n$. A major application of this theorem is in RSA cryptosystem.

Example 36   *1. Let $a = 12, n = 17$. Then $a^{33} \mod 17 = 12^{2 \times \phi(17)+1} \mod 17 \equiv 12^{2 \times 16+1} \mod 17 \equiv 12^1 \mod 17 = 12$.*

2. Let $a = 20, n = 77$, then $20^{62} \mod 77 \equiv (20^{\phi(77)} \mod 77)(20^2 \mod 77) \equiv 15 \mod 77$(Using Euler's theorem with $k = 1$).

We can see the application of Euler's theorem in finding the multiplicative inverses under composite modulo by the expression $a^{-1} \mod n \equiv a^{\phi(n)-1} \mod n$ (since $a^{\phi(n)} \mod n \equiv 1 \mod n$).

Example 37 *Let us do the following examples to find multiplicative inverse under composite modulo.*

1. *Find the value of $95^{-1} \mod 288$. We see that, $95^{-1} \mod 288 \equiv 95^{\phi(288)-1} \mod 288 \equiv 95^{96-1} \mod 288 \equiv 95^{95} \mod 288 \equiv 191 \mod 288$.*

2. *Find the value of $122^{-1} \mod 343$. We see that, $122^{-1} \mod 343 \equiv 122^{\phi(343)-1} \mod 343 \equiv 122^{294-1} \mod 343 \equiv 122^{294} \mod 343 \equiv 194 \mod 343$.*

3. *Find the value of $8^{-1} \mod 77$. We see that, $8^{-1} \mod 77 \equiv 8^{\phi(77)-1} \mod 77 \equiv 8^{60-1} \mod 77 \equiv 8^{59} \mod 77 \equiv 29 \mod 77$.*

4. *Find the value of $71^{-1} \mod 100$. We see that, $71^{-1} \mod 100 \equiv 71^{\phi(100)-1} \mod 100 \equiv 71^{40-1} \mod 100 \equiv 71^{39} \mod 100 \equiv 31 \mod 100$.*

3.4   Congruences

Let n be a positive integer. Consider any two integers a and b. Then, we say that a is congruent to b modulo n if n divides $a - b$ i.e $a = b + kn$ for some integer k. We denote this by $a \equiv b \mod n$ (also denoted as $a = b \mod n$). This relation is called congruence relation. The number n is called modulus for such a congruence relation. **Property of congruences**–For a fixed n, congruence relation follows the properties below:

1. $a \equiv a \mod n$.

2. $a \equiv b \mod n$ iff $a \mod n \equiv b \mod n$.

3. $a \equiv b \mod n \Rightarrow b \equiv a \mod n$.

4. $a \equiv b \mod n$ and $b \equiv c \mod n \Rightarrow a \equiv c \mod n$.

5. If $a \equiv c \mod n$ and $b \equiv d \mod n$, then $a+b \equiv c+d \mod n$ and $a.b \equiv c.d \mod n$.

6. If $a \equiv a \mod n$ then $gcd(a,n) = gcd(b,n)$.

7. $-x \mod n \equiv n - x \mod n$.

8. $y^{-1} \pmod{n}$ represents the inverse of $y \pmod{n}$ in $[1, n-1]$ (if exists) such that $yy^{-1} \pmod{n} \equiv 1$.

9. If $\gcd(a,n) = 1$, then $a^{-1} \mod n$ exists and can be obtained using Extended Euclidian Algorithm.

3.4.1 Solving Linear Congruence in Z_n

In cryptology, during the analysis of some cryptosystems, sometimes we need to solve a linear congruence equation. We state some basic theorems to arrive at the solution of a congruence equation. Their proofs can be obtained from the books on number theory [143, 6, 47].

Theorem 3.3
For an integer $n > 1$ and $d \neq 0$, if $ad \equiv bd \pmod{n}$, then $a \equiv b \pmod{\frac{n}{gcd(d,n)}}$.

Proof 3.3 Let $k = gcd(d,n)$. Then, $ad \equiv bd \pmod{n} \Rightarrow n|ad - bd \Rightarrow n|(a-b)d$
$\Rightarrow \frac{n}{k}|\frac{d}{k}(a-b)$, since $k = gcd(d,n) \Rightarrow \frac{n}{k}|(a-b) \Rightarrow a \equiv b \pmod{\frac{n}{k}}$.

Theorem 3.4
For integer $n > 1$, a necessary and sufficient condition that the congruence $ax \equiv b \mod n$ is solvable is that $gcd(a,n)$ divides b.

Remark 3.2 Any congruence equation $ax \equiv b \pmod{n}$ has $gcd(a,n)$ different solutions which are less than n. These solutions are given by $x + \frac{ni}{gcd(a,n)}$, where $i = 0, 1, 2...gcd(a,n) - 1$. If $gcd(a,n) = 1$, then the congruence equation has a unique solution.

Remark 3.3 If $f(x)$ is a polynomial over Z and $a \equiv b \pmod{n}$ for $a > b$, then $f(a) \equiv f(b) \pmod{n}$.

Remark 3.4 $ax \equiv ay \mod n$ if and only if $x \equiv y \mod \frac{n}{gcd(a,n)}$.

Example 38 *For,* $2x \equiv 5 \pmod{10}$, $gcd(2, 10) = 2$ *and* 2 *does not divide* 5. *Hence, no solution exists.*

Example 39 *For,* $6x \equiv 18 \pmod{36}$, $gcd(6, 36) = 6$ *and* 6 *divides* 18. *Hence, a solution* $x = 3$ *exists. The six different possible solutions are given by* $x = x + \frac{n\ i}{gcd(a,n)}$, *where* $i = 0, 1, 2, 3, 4, 5$. *i.e.* $x = 3, 9, 15, 21, 27$.

Example 40 *For,* $4x \equiv 2 \pmod{15}$, $gcd(4, 15) = 1$. *Hence, the solution is unique. The unique solution is given by* $x = 4^{-1}2 \mod 15 = 8$.

Remark 3.5 We can find the inverse of a matrix M in the modulo system Z_n just as we can find it in the real number system. The matrix M is invertible if the determinant of M ($det\ M$) is invertible in Z_n. The inverse is given by $M^{-1} = (det\ M)^{-1}adj(M)$, where $adj\ (M)$ denotes the adjoint of the matrix M. Here, all the operations are performed under modulo n.

3.4.2 Chinese Remainder Theorem (CRT)

This theorem is used to solve a set of congruent equations with one variable but relatively different prime moduli. Suppose, $m_1, ..., m_r$ are pairwise relatively prime positive integers and suppose $a_1, ..., a_r$ are integers. Then, the system of r congruences

$x \equiv a_i \pmod{m_i}, (1 \le i \le r)$ has a unique solution modulo $M = m_1 \times ... \times m_r$ which is given by $x = \sum a_i M_i b_i \mod M$, where $M_i = M/m_i$ and $b_i = M_i^{-1} \mod m_i$, for $1 \le i \le r$. See Algorithm 3.

Algorithm 3 (Chinese Remainder Theorem)

INPUT - Integers tuple $(m_1, m_2, ..., m_k)$, pair-wise disjoint; integers tuple $(a_1 \mod m_1, a_2 \mod m_2, ..., a_k \mod m_k)$.

OUTPUT - Integer $x < M = m_1 \times m_2 \times ... \times m_k$ satisfying $x = a_i \mod m_i \forall i = 1, 2, ..., k$.

1. $M \longleftarrow m_1 \times m_2 \times ... \times m_k$;
2. For (i from 1 to k) do
 a) $b_i \longleftarrow (M/m_i)^{-1} \mod m_i$; (by Extended Euclidean Algorithm)
 b) $c_i \longleftarrow b_i M/m_i$;
3. Return $(c_1 a_1 + c_2 a_2 + ... + c_k a_k) \mod M$.

Time Complexity of CRT: Time complexity of the Chinese Remainder Theorem is $\mathcal{O}(k(logM)^2)$.

Example 41 *Solve the simultaneous equations* $x \equiv 2 \mod 3$, $x \equiv 3 \mod 5$, $x \equiv 2 \mod 7$.

To solve the given congruence equations, follow the following steps-

1. *Step-1.* $M = 3 \times 5 \times 7 = 105$.

2. *Step-2.* $M_1 = \frac{105}{3} = 35, M_2 = \frac{105}{5} = 21, M_3 = \frac{105}{7} = 15$.

3. *Step-3. The inverses of* M_i *are* $b_1 = M_1^{-1} = 2$, $b_2 = M_2^{-1} = 1$, $b_3 = M_3^{-1} = 1$.

4. *Step-4. Finally, we compute* $x = (2 \times 35 \times 2 + 3 \times 21 \times 1 + 2 \times 15 \times 1) \mod 105 = 23 \mod 105$.

Applications of Chinese Remainder Theorem: This theorem is very useful in the field of cryptography. For example it is used in RSA cryptosystem to improve the efficiency of decryption procedure upto 4 times. Also, it is used for solving quadratic congruences and in represention of a large integer into several small integers.

3.5 Integer Factorization Problem

Through the fundamental theorem of arithemetic, we know that every number greater than 1 can be uniquely expressed as a product of powers of distinct primes. Now, the question is, how quickly can we factorize an integer. There are so many algorithms for integer factorization. But all of those are not polynomial time algorithms. Our interest is to know that if there exists any algorithm that can factor any integer n in polynomial time. To answer this question, P.W. Shor [158] devised a polynomial time algorithm for factorization but with using quantum computers. In this section, we study some basic factorization algorithms.

3.5.1 Trial Division Method

The first algorithm for factorization is trial division method. The trial division method for factoring a number is simply to find factors by trial and error approach. In this method for given an input n, we pick all the possible divisors d between 2 and $\sqrt{n}$, check if d divides n and if it does, then d is a divisor. Look at Algorithm 4 below.

Algorithm 4 Trial Division Factorization(n)

INPUT- An integer n.
OUTPUT- All prime factors of n.

$d \leftarrow 2$
while $(d \leq \sqrt{n})$
$\{$
$d \leftarrow 2$
while $(n \mod d = 0)$
$\{$
output a (output factors one by one)
$n = n/d$
$\}$
$d \leftarrow a + 1$
$\}$
if $(n > 1)$ output n (n has no more factors)

Explanation of Algorithm 4. If n is a composite integer $n = p \times q$ (say), then one of the factors must be less than $\sqrt{n}$. The trial factors should not be too distant from $\sqrt{n}$.

Example 42 *Let $n = 12$ which is to be factored. The numbers dividing n are $2, 3, 4, 6, 12$. We check only for the numbers $< \sqrt{12}$ i.e. $2, 3$. Running the above algorithm, we find $12 = 2 \times 2 \times 3$. Clearly 2 and 3 are the divisors of other factors like 4, 6 and 12. Therefore we do not need to check.*

3.5.2 Fermat's Method

Fermat's factorization method, named after the French mathematician Pierre de Fermat (1607–1665), is based on the representation of an odd integer as the difference of two squares. It proves to be a fast method when the factors are closer to the root of the number. When the factors aren't close to the root of the number, the method proves to be quite slow, perhaps as slow as the trial division method. Let us look at Algorithm 5.

Explanation of Algorithm 5. If $n = a \times b$ is an odd integer, then the factors a and b both are also odd integers. We can express $n = (\frac{a+b}{2})^2 - (\frac{a-b}{2})^2 = x^2 - y^2$, where $x = \frac{a+b}{2}$ and $y = \frac{a-b}{2}$. This method uses the fact that if we can find x and y such that $n = x^2 - y^2$, then we have $n = x^2 - y^2 = (x+y)(x-y) = a \times b$ with $a = x+y, b = x-y$. This method starts from smallest integer greater than $x = \sqrt{n}$ and tries to find y such that $y^2 = x^2 - n$ holds. In each iteration, we check that $x^2 - n$ is a perfect square. If it is so, the work is done.

Algorithm 5 Fermat Factorization (n)

INPUT- An odd integer n.
OUTPUT- All prime factors of n.
{
$x \leftarrow \sqrt{n}$
while $(x < n)$
{
$w \leftarrow x^2 - n$
if (w is perfect square) $y \leftarrow \sqrt{n}$; $a \leftarrow x + y$; $b \leftarrow x - y$; return a and b
$x \leftarrow x + 1$
}
}

Example 43 *Suppose* $n = 667$ *is to be factored. Then* $x = 26 > \sqrt{667}$. $y^2 = x^2 - n = 26^2 - 667 = 676 - 667 = 9$ *(a perfect square).* $\Rightarrow a = x + y = 26 + 3 = 29, b = x - y = 26 - 3 = 23 \Rightarrow 667 = 29 \times 23$.

Example 44 *Suppose* $n = 5959$ *is to be factored. Then we would first try with for* $x = 78 = \lceil \sqrt{5959} \rceil$, *then* $y^2 = 78^2 - 5959 = 125$. *Since 125 is not a square, we would again try by increasing the value of* x *by 1. In the second attempt, we find* $y = 282$, *but 282 is again not a perfect square. The third trial produces the number 441, which is a perfect square of 21. Hence,* $x = 80$, $y = 21$, *and the factors of 5959 are* $x - y = 59$ *and* $x + y = 101$.

Remark 3.6 Suppose n has more than two prime factors. This procedure first finds the factorization with the least values of x and y. That is, $x + y$ is the smallest factor $\geq \sqrt{n}$. And so $x - y = \frac{n}{x+y}$ is the largest factor $\leq \sqrt{n}$. If the procedure finds $n = 1 \times n$, this shows that n is prime.

3.5.3 Pollard's $p - 1$ Method

Pollard's $p - 1$ algorithm is named after the British mathematician John Pollard. This algorithm is only suitable for integers with specific types of factors. It finds those factors p, for which $p - 1$ is powersmooth.
Main Idea. Let $n = p \times q$. Consider an integer m s.t. $(p - 1)|m$ but $(q - 1) \nmid m$. Now choose an $x_n \in Z_n^\star$ at random and compute $y = (x_n^m - 1) \mod n$. Then according to the Chinese Remainder Theorem, we can find $x_p \in Z_p^\star$ and $x_q \in Z_q^\star$ such that $y = (x_p^m - 1 \mod p, x_q^m - 1 \mod q)$. Using the fact that $(p - 1)|m$ and the Fermat's Little Theorem, we get $y = (0, x_q^m - 1 \mod q)$. Then clearly $p|y$ but $q \nmid y$. Thus, $gcd(y, n) = p$. Subsequently, by computing one gcd we can obtain the prime factor of n. Let us look at Algorithm 6.

Algorithm 6 Pollard's $p-1$ Factorization

INPUT- A composite integer n and B.

OUTPUT- A non-trivial factor of n or failure.

```
{
a ← 2
e ← 2
while (e ≤ B)
{
a ← aᵉ  mod n
e ← e + 1
}
d ← gcd (a − 1, n)
if 1 < p < n
return d
else
return failure
}
```

Explanation of Algorithm 6. In this algorithm, we take two inputs. The first is an odd integer n which is to be factored and a pre-defined value, a bound B. Now, if p is a prime divisor of n and $q \le B$ for every prime power $q|p-1$, then , $p-1|B!$. At the end of the loop, we have $a \equiv 2^{B!} \mod n$. Now, since $p|n \Rightarrow a \equiv 2^{B!} \mod p$, hence by Fermat's Theorem, $2^{p-1} \equiv 1 \mod p$ and $p-1|B! \Rightarrow a \equiv 1 \mod p \Rightarrow a|p-1$. We also have that $p|n \Rightarrow p|gcd(a-1,n)$. Suppose, $gcd(a-1,n) = d$, then d and $\frac{n}{d}$ are the factors of n.

The time complexity of this algorithm is $\mathcal{O}(B \ log \ B \ (log \ n)^2 + (logn)^3)$. The drawback of this algorithm is that p should be a smooth prime (n is to have a prime factor p such that $p-1$ has only small prime factors) for solving factorization problem successfully by this algorithm. Let us do the following examples.

Example 45 *Let* $n = 15770708441$ *and* $B = 180$. *By Pollard's* $p-1$ *algorithm, we find* $a = 11620221425$, $d = 135979$ *i.e.* $n = 15770708441 = 135979 \times 115979$. *The prime factors of* $p-1 = 135978 = 2 \times 3 \times 131 \times 173$. *We can choose* $B \ge 173$. *Since the prime factors of* $q-1 = 115978 = 2 \times 103 \times 563$, *if we choose* B *between 173 and 562, we always get the factor 135979.*

Example 46 *To factor the value n = 6994241 using Pollard's p − 1 algorithm, see the calculation given in the following table.*

e	a	a^e mod n	d
2	2	4	1
3	4	64	1
4	64	2788734	1
5	2788734	3834705	1
6	3834705	513770	1
7	513770	443653	3361

Thus, for B ≥ 7 we get a prime factor 3361 of n. Finally, we get n = 3361 × 2081. The prime factor of 3360 = 2^5 × 3 × 5 × 7, hence, it is a 7 smooth number. And the prime factors of 2080 = 2^5 × 5 × 13, hence it is a 13 smooth number. So if we choose the bound B between 7 and 12, we get the factor 3361.

3.5.4 Pollard's Rho Method

This method is given by the British mathematician John M Pollard in 1975. The algorithm can be used to compute factors of any arbitrary integer $n = p \times q$. In this approach, we find two distinct elements $x, y \in Z_n^*$ such that x mod $p = y$ mod p. It is clear that for this pair $gcd(x - y, n) = p$. Therefore, computing the gcd, one can extract a non-trivial prime factor of n. See the Algorithm 7.

Algorithm 7 Pollard's Rho Method

(n is the number to be factored)

$x \leftarrow 2$

$y \leftarrow 2$

$p \leftarrow 1$

while ($p = 1$)

{

$x \leftarrow f(x)$ mod n

$y \leftarrow f(f(y)$ mod n) mod n

$p \leftarrow gcd(x - y, n)$

}

return p (if $p = n$, then program is failed)

}

Explanation of Algorithm 7. Suppose p is the smallest prime divisor of n. Assume there exist two integers $x, y \in Z_n$ such that $x \neq y$ but $x \equiv y$ mod p. Then, $p \leq gcd(x - y, n) < n$. Hence, we obtained a non-trivial factor of n by computing $gcd(x - y, n)$.

This algorithm will terminate successfully if and only if there exists at least one collision i.e. $x \neq y$ but $x \equiv y \mod p$. To find the appropriate pair (x, y) we proceed as follows. Let p be a prime factor of a composite integer n. Pollard's Rho Algorithm for factoring n actually attempts to find duplicate in the sequence of integers $x_0, x_1, x_2 \ldots$ defined by $x_0 = a$ (a constant 2 say), $x_{i+1} = f(x_i)$ ($= x_i^2 + 1$, say). If $x_i = x_j \pmod{p}$ then $f(x_i) = f(x_j) \mod p$, i.e. $x_{i+1} = x_{j+1} \mod p$ and hence $x_{i+\delta} = x_{j+\delta} \mod p$, in particular we have $x_i = x_{2i} \mod p$ but $x_i \neq x_{2i} \mod n$. This is based on Theorem 3.5.

Theorem 3.5
Let $(x_k \mod n)_{k \geq 0}$ be a pseudo random sequence in Z_n. If there exist indices $i < j$ such that $x_i \equiv x_j \mod p$, then there also exists an index $i \leq r < j$ such that $x_r \equiv x_{2r} \mod p$.

The probability of getting at least one collision can be viewed by using birthday paradox which is equal to $1/2$. Assuming that the function $f(x) = x^2 + 1$ behaves like a random function, the method requires $\mathcal{O}(\sqrt{p})$ arithmetic operations to find a non-trivial factor p of n. In general, since $p < \sqrt{n}$, the expected time to find non-trivial factor of n is $\mathcal{O}(n^{\frac{1}{4}})$. Thus, the bit operations complexity is $\mathcal{O}(2^{\frac{\log n}{4}})$.

3.5.5 Quadratic Sieve

Quadratic field sieve (proposed in 1981 by the American number theorist C. B. Pomerance) is an algorithm for solving factorization problem. It is known as the second fastest algorithm after number field sieve for integers under 100 decimal digits. Fermat's method is based on to search a number x such that "$x^2 \mod n$" is a square. But these x are very difficult to find. Quadratic sieve computes "$x^2 \mod n$" for several x and then finds a subset of these numbers, whose product is a square that gives a congruence of squares. Hence, this method uses essentially a sieving procedure to find the value of "$x^2 \mod n$". Its running time is $\mathcal{O}(e^c)$, where $c \approx (ln\, n\, ln\, ln\, n)^{\frac{1}{2}}$. Consider the following example of quadratic sieve.

Example 47 *Consider $n = 3837523$ is a number to be factored. Then, we see that*
$9398^2 \equiv 5^5.19 \mod 3837523$
$19095^2 \equiv 2^2.5.11.13.19 \mod 3837523$
$1964^2 \equiv 3^2.13^3 \mod 3837523$
$17078^2 \equiv 2^6.3^2.11 \mod 3837523$
If we multiply all these together, we get
$(9398.19095.1964.17078)^2 \equiv (2^4.3^2.5^3.11.13^2.19)^2 \mod 3837523$
which gives $2230387^2 \equiv 2586705^2 \mod 3837523$
But, $2230387 \neq \pm 2586705 \mod 3837523.$
Now, compute $gcd(2230387 - 2586705, 3837523) = 1093$. *Therefore,* $3837523 = 1093 \times 3511$.

3.5.6 Number Field Sieve

Number field sieve method was developed by John Pollard in 1988, and is considered to be an extremely quick factorization method which had been used for factoring RSA-130. This method uses a sieving procedure in an algebraic ring structure to find $x^2 \equiv y^2 \mod n$. Its running time is $\mathcal{O}(e^c)$, where $c \approx 2(ln\ n)^{\frac{1}{3}}(ln\ ln\ n)^{\frac{2}{3}}$. This method reduces the exponent over quadratic sieve method.

Example 48 *Assume that there is a computer that can perform 2^{30} (almost 1 billion) bit operations per second. What is the approximate time required for this computer to factor an integer of 100 decimal digits using number field sieve method? Since a number with 100 decimal digits has almost 300 bits ($n = 2^{300}$), $ln(2^{300}) = 207$ and $lnln(2^{300}) = 5$, then, $(207)^{\frac{1}{3}} \times (5)^{\frac{2}{3}} = 6 \times 3 \approx 18$. Hence, the time is $\frac{e^{18}}{2^{30}} \approx 6$ seconds.*

The complete explanation of both the methods, Quadratic sieve and Number field sieve, is out of the scope of this book. The interested readers can refer the book [176].

3.6 Primality Testing

The process or algorithm to test 'whether the given number is prime or not' is called primality testing. Many primality testing algorithms are available in the literature. In this section we will discuss some basic primality testing techniques.

3.6.1 Sieve of Eratosthenes

The Greek Mathematician Eratosthenes derived a method to find all primes less than a given number n. Let us see how this method works. Suppose, we want to find all primes less than 100. Since $\sqrt{100} = 10$ and the primes less than 10 are 2, 3, 5 and 7, we first write all the numbers 1,2,...100 in a table of 10 by 10 squares. Then, we cross all multiples of 2, 3, 5, and 7. The remaining numbers are primes less than 100. Using the Sieve of Eratosthenes, we can generate all primes with given upper bound. We can also use this method for primality testing but for that we would need a bigger table for a large number. In cryptography, many times we need to generate a prime number of a given size. For this, we first choose a number of given size and then apply some primality testing algorithm to decide whether the chosen number is prime or not and would further proceed accordingly. Testing primality belongs to the class of decisional problems. As we would also explain in Chapter 5, a decisional problem is a problem in which a question is to be answered in either "YES" or "NO". There are two different types of primality testing algorithms, deterministic and randomized (probabilistic).

We will first elucidate the deterministic algorithm for primality testing.

3.6.2 Divisibility Algorithm

This is the most elementary deterministic method of primality testing. It is similar to the Seive of Eratosthenes. It is based on the concept that if the number n is composite then one of its prime factors must be less than $\sqrt{n}$. In order to test whether the given number is prime or not, we divide the given number n by all primes less than $\sqrt{n}$.

Algorithm 8 Divisibility test (n)

---

n is the number to test for primality
INPUT- A number n;
OUTPUT- Composite or Prime;
$r \leftarrow 2$
while $(r < \sqrt{n})$
{
if $(r|n)$
Return ("*a* **composite**")

$r \leftarrow r+1$
}
Return ("*a* **prime**")

This is an exponential time algorithm. The time complexity of this algorithm is $2^{\frac{n_b}{2}}$, where n_b is the bit size of the given integer n. In big-$\mathcal{O}$ notation the time complexity is given by $\mathcal{O}(2^{n_b})$. Thus, if we assume that each arithemetic operation uses only one bit operation then the total number of bit operations required in this algorithm is $2^{\frac{n_b}{2}}$.

Example 49 *Assuming n as a 200 bit number, the number of bit operations needed to run the divisibility algorithm needs 2^{100} bit operations. On a computer capable of computing 2^{30} bit operations per second, the algorithm needs 2^{70} seconds to complete the testing successfully.*

3.6.3 AKS Algorithm

AKS algorithm is Agrawal-Kayal-Saxena primality testing algorithm. It is a deterministic algorithm. This can be used to verify the primality of any given number. If n_b is the size of the input integer its complexity is $\mathcal{O}(log_2 n_b)^{12}$. This is a polynomial time algorithm but is not a practically efficient algorithm due to its higher order.

Example 50 *Assuming n as a 200 bit number, the number of bit operations needed to run the AKS algorithm is only* $(log_2 200)^{12} = 39,547,615,483$. *On a computer capable of computing 1 billion bit operations per second, the algorithm needs only 40 seconds.*

In the rest of the chapter, we will discuss only randomized algorithm.

3.6.4 Fermat Test

Fermat test is based on Fermat's Little Theorem. If n is prime, then $a^{n-1} \equiv 1 \mod n$. This theorem can be used to determine whether a given positive integer is composite or not. However, it cannot prove that a given number is prime or not. Hence, it is also called Fermat's compositeness test. To apply this test for a given positive integer n, we choose a positive integer $a < (n-1)$ and compute $y = a^{n-1} \mod n$, if $y \neq 1$ for any $a < n$, then we conclude that the given number is composite. But, if $y = 1$, then we repeat the process for a different $a < (n-1)$ until we get $y \neq 1$.

Example 51 *Consider* $n = 341$, *if we apply Fermat test with* $n = 341$ *and* $a = 2$, *then we get* $a^{340} \equiv 1 \mod 341$. *Again, if we repeat the test for* $a = 3$, *then we see that* $3^{340} \equiv 56 \mod 341$. *Thus, the given number is a composite number. In fact,* $341 = 11 \times 31$.

In Fermat test, if $a^{n-1} \equiv 1 \mod n$ where $a < n$, then we say that the number n is *pseudoprime* to the base a. But, sometimes if n is a composite number then also it is possible that $a^{n-1} \equiv 1 \mod n$ for all a relatively prime to n i.e n is pseudoprime to the base a for all integers with gcd $(a,n) = 1$. Such type of integers are called *Carmichael Number* . The smallest Carmichael number is 561.

Example 52 *Consider the number* $n = 561$. *By using base* 2, *we see that* $2^{561-1} \equiv 1 \mod n$. *This number follows the Fermat test but is not prime since* $561 = 3 \times 11 \times 17$.

Due to the existence of Carmichael number, the Fermat test for primality is a NO-based Monte Carlo algorithm. Please refer to definition 5.23 in chapter 5.

3.6.5 Miller-Rabin Algorithm

This algorithm can be used to prove the compositeness of n if the number n is composite. Unlike Fermat test, there are no Carmichael numbers for Miller-Rabin test.

Explanation of Algorithm 9. This is a polynomial time algorithm. Its running time complexity is $\mathcal{O}(log\ n)^3$. If this algorithm terminates with an answer 'yes', then it is correct and the number is truly composite. The reason is as follows. If n is prime, then the order of Z_n^* is $n - 1 = 2^k m$. If we define $T = a^m$, then its order must be a power of 2, i.e. in the form of 2^r. Now there are two cases: first, if its order is $t = 1 = 2^0$, then $a^m \equiv 1 \mod n$; second, if its order is $t > 1$, then $t = 2^i$ with $1 \leq i \leq k$, thus $T^{2^i} \equiv 1 \mod n$. Since n is prime and order of T is 2^i, we have $T^{2^{i-1}} \equiv -1$

Algorithm 9 Miller-Rabin Test(*n*)

Write $n - 1 = 2^k m$, (where *m* is odd)
Choose at random a value *a*, such that $1 < a \leq n - 1$
$T \to a^m \mod n$
if $T \equiv 1 \pmod{n}$
then return "*n* is prime"
For $i \leftarrow 0$ to $k - 1$

do $\begin{cases} \quad \textbf{if } T \equiv -1 \pmod{n} \\ \textbf{then return}\text{"}n \text{ is prime"} \\ \quad \textbf{else } T \leftarrow T^2 \mod n \end{cases}$

return "*n* is composite"
(**Comment**: *a* is called witness of the compositeness of *n*)

---

mod *n*, i.e. $a^{2^{i-1}m} \equiv -1 \mod n$. Thus, we have the statement- "If *n* is prime and $gcd(a, n) = 1$ then either $a^m \equiv 1 \mod n$ or there exists $i \in \{0, 1, 2, 3...k\}$ such that $a^{2^i m} \equiv -1 \mod n$".

Let us discuss the correctness of the theorem. We want to check if the theorem outputs correct decision. For this, suppose the algorithm outputs 'composite' for a 'prime' number *n*. Since the algorithm tells that *n* is composite $\Rightarrow a^m \not\equiv 1 \mod n$, in each iteration of the loop, when $T = a^m$ is squared $\Rightarrow a^{2m} \not\equiv -1 \mod n$, ..., $a^{2^{k-1}m} \not\equiv -1 \mod n$. Since, we have assumed that *n* is prime $\Rightarrow a^{n-1} \equiv 1 \mod n \Rightarrow a^{2^k m} \equiv 1 \mod n \Rightarrow a^{2^{k-1}m}$ is a square root of 1 mod *n*. Since *n* is prime, therefore, there exist only two square roots ± 1 of 1 mod *n*. We know that, $a^{2^{k-1}m} \not\equiv -1 \mod n \Rightarrow a^{2^{k-1}m} \equiv 1 \mod n$, so, $a^{2^{k-1}m}$ should be the square root of 1 mod *n*. By continuing this process, we get $a^m \equiv 1 \mod n \Rightarrow n$ is prime. But, by algorithm, *n* is composite which is a contradiction. Hence, if *n* is prime, then this algorithm can never produce the output '*n* is composite'.

Example 53 *Suppose we want to check whether $n = 221$ is prime or not. For this, since $n - 1 = 220 = 2^2.55$, so here $k = 2, m = 55$. Now, we randomly select a number $1 < a < n - 1$ let $a = 174$. Now, $a^{2^0.m} \mod n = 174^{55} \mod 221 = 47 \neq 1$. $a^{2^1.m} \mod n = 174^{110} \mod 221 = 220 = -1$. Since, $220 = -1 \mod 221$. Therefore, either 221 is prime or 174 is a strong liar for 221. Now, we try for another $a = 137$. Here $a^{2^0.m} \mod n = 137^{55} \mod 221 = 188 \neq 1$. $a^{2^1.m} \mod n = 137^{110} \mod 221 = 205 \neq n - 1$. Hence, 137 is a witness of the compositeness of 221.*

Remark 3.7 If the algorithm terminates with answer NO i.e. "*n* is prime", the chances of the number being prime are greater than 75%. If *n* is composite, then it passes the strong primality test to the base *a* with less than 25% for all *a* in the range $1 < a < n - 1$. If the primality test is performed for *k* different randomly chosen values of a, and n passes the primality test for all of these a, then we can say that there is at least $1 - \frac{1}{4^k}$ probability that n is prime.

The success probability of Miller-Rabin test can be estimated by the following theorem.

Theorem 3.6
If $n \geq 3$ is an odd composite number, then the set $\{1, 2, ..., n-1\}$ contains at most $\frac{n-1}{4}$ numbers that are prime to n and not witness for the compositeness of n.

Due to space limitation we omit the proof of the theorem.

3.7 Quadratic Congruence

The equations of the form $a_2 x^2 + a_1 x + a_0 \equiv 0 \bmod n$ are called quadratic congruence modulo n. We are concentrating on the case when a_1 is 0 and a_2 is 1 i.e. equations of the form $x^2 \equiv a \bmod n$.

Example 54 $x^2 \equiv 3 \mod 11$ *has two solutions,* $x \equiv 5 \mod 11$ *and* $x \equiv -5 \mod 11$. *Further, since* $-5 \mod 11 \equiv 6 \mod 11$, *therefore the two solutions are* $x = 5$ *and* 6.

Example 55 $x^2 \equiv 2 \mod 11$ *has no solution. No integer x satisfies the relation* $x^2 \equiv 3 \mod 11$.

3.7.1 Quadratic Residue or Non-Residue

Let $n > 1$ be an integer. For $a \in Z_n^\star$, a is called quadratic residue mod n if $x^2 = a$ (mod n) for some $x \in Z_n^\star$, otherwise it is called quadratic non-residue modulo n. The set of quadratic residue modulo n is denoted by QR_n and the set of quadratic non-residues modulo n is denoted by QNR_n.

Theorem 3.7
Let p be a prime number, then $QR_p = \{x^2 \pmod{p} | 0 < x \leq \frac{p-1}{2}\}$. There are precisely $\frac{p-1}{2}$ elements which are QR_p and $\frac{p-1}{2}$ elements which are QNR_p i.e. $Z_p^\star$ can be divided into two equal sized subsets QR_P and QNR_P.

Example 56 *Let us consider $Z_5^\star$. In $Z_5^\star$, $1^2 = 1, 2^2 = 4, 3^2 = 4, 4^2 = 1$. Clearly, 1 and 4 are quadratic residues under modulo 5 and 2 and 3 are not quadratic residues under modulo 5.*

Theorem 3.8
Let n be a composite integer with $k > 1$ distinct prime factors. Then exactly $\frac{1}{2^k}$ fractions of elements of $Z_n^\star$ are QR_n.

Definition 3.11   Quadratic Residuosity Problem (QRP): Quadratic residuosity problem is to decide whether a given number a is quadratic residue for given modulus n or not.

The important applications of QRP in cryptography is in the design of Goldwasser-Micali cryptosystem, Rabin cryptosystem, Cock's identity-based encryption, and corresponding signature schemes. Using Euler's criterion, we can test whether an integer is quadratic residue modulo prime or not. Till date, for a composite n of unknown factorization, no algorithm is known to be able to decide quadratic residuosity modulo n in polynomial time in the size of n.

Theorem 3.9
Euler Criterion for Quadratic residuosity. *If p is prime and $a \in Z_p^\star$, then $a \in QR_p$ if and only if $a^{\frac{p-1}{2}} \equiv 1 \pmod{p}$.*

Proof 3.9 Let $a \in QR_p$. Then $\exists x \in Z_p^\star$ such that $x^2 \equiv a \pmod{p}$, hence $a^{\frac{p-1}{2}}$ mod $p \equiv x^{p-1} \equiv 1 \mod p$.
Conversely, assume that $a^{\frac{p-1}{2}} \equiv 1 \mod p$. Since $Z_p^\star$ is a cyclic group (p is prime). Let g be a primitive root modulo p. Then $Z_p^\star = <g>$. For given $a \in Z_p^\star$, there exists $i; 1 \leq i \leq p-1$ such that $a \equiv g^i \pmod{p} \Rightarrow g^{i\frac{p-1}{2}} \equiv a^{\frac{p-1}{2}} \mod p \equiv 1 \pmod{p}$ (by assumption). By the property of primitive root, $(p-1)$ must divide $i(\frac{p-1}{2})$. Therefore, i must be even i.e. $i = 2j \Rightarrow a = (g^j)^2 \mod p. \Rightarrow a \in QR_p$.

Let us do following examples to apply the above theorem.

Example 57 *For $a = 17, p = 19$, $17^{\frac{19-1}{2}} \mod 19 = 17^9 \mod 19 = 1 \mod 19$. Hence, 17 is quadratic residue under modulo* 19.

Example 58 *For $a = 17, p = 11$, $17^{\frac{11-1}{2}} \mod 11 = 17^5 \mod 11 = -1 \mod 11$. Hence, 17 is quadratic non-residue under modulo* 11.

3.7.2   Legendre Symbol and Jacobi Symbol

The Legendre Symbol is a useful tool to test whether or not an integer a is a quadratic residue modulo a prime p. Euler's criterion for testing quadratic residuosity involves evaluation modulo exponentiation which is quite computation intensive. Quadratic residuosity can be tested by a much faster algorithm, an algorithm that is based on Legendre-Jacobi Symbol.

Legendre Symbol: Let a be an integer and let p be an odd prime. The Legendre symbol $\left(\frac{a}{p}\right)$ is defined as follows:

$$\left(\frac{a}{p}\right) = \begin{cases} 1: & if & a \in QR_p \\ -1: & if & a \in QNR_p \\ 0: & if & p|a \end{cases}$$

Example 59 *Consider the group* $Z_5^* = \{1,2,3,4\}$. *Since* $1^1 = 1, 2^2 = 4, 3^2 = 4, 4^2 = 1$, *we have* $QR_5 = \{1,4\}$ *and* $QNR_5 = \{2,3\}$. *Thus, we have the Legendre Symbol-* $\left(\frac{1}{5}\right) = 1, \left(\frac{2}{5}\right) = -1, \left(\frac{3}{5}\right) = -1, \left(\frac{4}{5}\right) = 1, \left(\frac{5}{5}\right) = 0$.

Remark 3.8 If p is prime and $a,b \in Z$ then, $\left(\frac{a}{p}\right) = a^{\frac{p-1}{2}}$ mod p, $\left(\frac{1}{p}\right) = 1$ and $\left(\frac{-1}{p}\right) = (-1)^{\frac{p-1}{2}}$. Hence, $-1 \in QR_p$ if $p \equiv 1$ (mod 4) and $-1 \in QNR_p$ if $p \equiv 3$ (mod 4)

Jacobi Symbol: Let m be an odd positive integer with prime factorization $m = \prod p_i^{e_i}$ and let a be an integer. The Jacobi Symbol $\left(\frac{a}{m}\right)$ is defined by

$$\left(\frac{a}{m}\right) = \prod \left(\frac{a}{p_i}\right)^{e_i}$$

where the symbols on the right hand side are Legendre Symbols. In the rest of this chapter, we denote the Jacobi Symbol by $\left(\frac{a}{n}\right)$. It is easily understood that if n is prime then the Jacobi Symbol is the Legendre Symbol.

Properties of Jacobi Symbol: Let m, n be the positive integers and let $a,b \in Z$. Then, the Jacobi Symbol has the following properties-

1. $\left(\frac{1}{n}\right) = 1$.

2. $\left(\frac{ab}{n}\right) = \left(\frac{a}{n}\right)\left(\frac{b}{n}\right)$.

3. $\left(\frac{a}{mn}\right) = \left(\frac{a}{m}\right)\left(\frac{a}{n}\right)$.

4. $a \equiv b$ mod n implies $\left(\frac{a}{n}\right) = \left(\frac{b}{n}\right)$.

5. If n is an odd integer then $\left(\frac{-1}{n}\right) = (-1)^{\frac{n-1}{2}}$.

6. $\left(\frac{2}{p}\right) = (-1)^{\frac{p^2-1}{8}}$. Hence, $\left(\frac{2}{p}\right) = 1$ if $p \equiv \pm 1$ (mod 8) and $\left(\frac{2}{p}\right) = -1$ if $p \equiv \pm 3$ (mod 8).

7. Law of Reciprocity - If m, n are odd numbers and $gcd(n,m) = 1$ then, $\left(\frac{p}{q}\right) = (-1)^{\frac{p-1}{2}\frac{q-1}{2}}\left(\frac{q}{p}\right)$. Hence, $\left(\frac{p}{q}\right) = -\left(\frac{q}{p}\right)$ if $p \equiv q \equiv 3$ (mod 4) and $\left(\frac{p}{q}\right) = \left(\frac{q}{p}\right)$ otherwise.

Due to space limitation, we are omiting the proof of these properties. The proof of the above properties can be obtained from any of the book on number theory viz.[143].

Algorithm 10 Algorithm for Legendre and Jacobi Symbols

INPUT - odd integer $n > 2$, integer $x \in Z_n^*$

OUTPUT - $\left(\frac{x}{n}\right)$

Jacobi $(x;n)$

1. If $(x = 1)$ return (1);

2. If $(2|x)$

a) If $\left(\left(2\left|\frac{n^2-1}{8}\right.\right)\right.$ return $\left(Jacobi\left(\frac{x}{2},n\right)\right)$;

b) return $\left(-Jacobi\left(\frac{x}{2},n\right)\right)$;

(now x is odd)

3). If $\left(2\left|\frac{(x-1)(n-1)}{4}\right.\right)$ return (Jacobi $(n \mod x, x)$);

4) return (- Jacobi $(n \mod x, x)$).

Based on the above mentioned properties, the following algorithm i.e. Algorithm 10 is designed to find Jacobi/Legendre Symbol.

Time Complexity: In Algorithm 10, each recursive call of the function Jacobi(;) will cause either the first input value being divided by 2 or the second input value being reduced under modulo the first value. Therefore, the process can be repeated to $log_2 n$ steps to reach the terminating condition. Thus, in general, because each modulo operation costs $\mathcal{O}(logn)^2$ time and we are repeating the modulo operation upto maximum $\mathcal{O}(log n)$ time, this algorithm computes Jacobi($x;n$) in $\mathcal{O}(logn)^3$ time hence the time complexity of Algorithm 10 is $\mathcal{O}(log n)^3$.

Remark 3.9 During the evaluation of Jacobi($x;n$) using Algorithm 10, it is not required to know the prime factorization of n. This is a very important property which has a wide application in PKC e.g. in Goldwasser-Micali cryptosystem, Rabin Cryptosystem, Cock's identity-based encryption etc.

Remark 3.10 For composite n, note that if the Jacobi Symbol $\left(\frac{a}{n}\right) = 1$, it does not mean that $a \in QR_n$.

If the prime factorization of n is known, then using the following theorem we can test whether a given $x \in Z_n^*$ is in QR_n or not.

Theorem 3.10
Let $n = p_1^{e_1} p_2^{e_2}...p_k^{e_k}$ be a composite number with its complete prime factorization. Then, $x \in QR_n$ if and only if $x \pmod{p_i^{e_i}} \in QR_{p_i^{e_i}}$ and hence, if and only if $x \pmod{p_i} \in QR_{p_i}$ for prime p_i with $i = 1, 2, ...k$.

Example 60 *Let us test whether 384 is quadratic non-residue modulo 443. Here, $x = 383$ and $n = 443$. Using Algorithm 10 we can calculate the Jacobi Symbol as*

Jacobi(383,443) = - Jacobi(192,443) $\{2$ *divides x but not* $\frac{n^2-1}{8}\}$
= Jacobi(96,443)
=-Jacobi(48,443)
=Jacobi(24,443)
=- Jacobi(12,443)
=Jacobi(6,443)
= - Jacobi(3,443)
= Jacobi(2,3) $\{x$ *is odd and* 2 *does not divide* $(x-1)(n-1)/4\}$
= Jacobi(2,3) = - Jacobi(1,3) = -1
Since 443 is prime, 384 is quadratic non-residue modulo 443 i.e $x \in QNR_{443}$.

By the Legendre Symbol, we can check whether the given number is quadratic residue or not. If $x \in QR_p$, then how would we calculate its square root? Using the following Algorithm 11, we can compute the square root modulo a prime.

Algorithm 11 Computing Square Root Modulo a Prime

INPUT - Prime p satisfying $p \equiv 3,5,7 \pmod{p}$
OUTPUT - A square root of $a \pmod{p}$
1. If $(p \equiv 3,7 \bmod 8)$ return $(a^{(p+1)/4} \pmod{p})$;
(below $p \equiv 5 \bmod 8$)
2. If $(a^{(p-1)/4} \equiv 1 \bmod p$ return $(a^{(p+3)/8} \bmod p))$
3. Return $((4a)^{(p+3)/8}/2 \pmod{p})$

Time Complexity: The time complexity of the Algorithm 11 is $\mathcal{O}(log p)^3$.

If the prime factorization of a composite number n is known, then using the above algorithm on each prime factor of n and then subsequently using the Chinese Remainder Theorem, we can compute the square root of x under composite modulo. See the following Algorithm 12.

Algorithm 12 Square root modulo composite

INPUT - Prime p,q with $n = pq$; integer $y \in QR_n$
OUTPUT - A square root of $y \pmod{n}$.
1. $x_p \longleftarrow \sqrt{y} \bmod p$;(*Applying algorithm for prime*)
$x_q \longleftarrow \sqrt{y} \bmod q$;
2. Return $((q^{-1} \bmod p) qx_p + (p^{-1} \bmod q) px_q) \bmod n$.
(*applying CRT*)

Time Complexity: The first two steps to compute square root under modulo p and modulo q requires $\mathcal{O}(log\,p)^3$ and $\mathcal{O}(log\,q)^3$ respectively and the application of Chinese Remainder Theorem requires $\mathcal{O}2(log\ n)^2$. Hence, the total time complexity of Algorithm 12 is $(\mathcal{O}(log\ p)^3 + \mathcal{O}(log\ q)^3 + \mathcal{O}(2(log\ n)^2)) \approx \mathcal{O}(log\ n)^3$.

Example 61 *Let us solve* $x^2 \equiv 36 \mod 77$. *Here,* $77 = 7 \times 11$. *So, we have* $x^2 \equiv 36 \mod 7 \equiv 1 \mod 7$ *and* $x^2 \equiv 36 \mod 11 \equiv 3 \mod 11$. *The roots satisfying these equations are* $x \equiv +1 \pmod 7$, $x \equiv -1 \pmod 7$, $x \equiv +5 \pmod{11}$, *and* $x \equiv -5 \pmod{11}$. *Now, the following set of equations can be made from the following equations:*

$x \equiv +1 \mod 7$ *and* $x \equiv +5 \mod 11$
$x \equiv +1 \mod 7$ *and* $x \equiv -5 \mod 11$
$x \equiv -1 \mod 7$ *and* $x \equiv +5 \mod 11$
$x \equiv -1 \mod 7$ *and* $x \equiv -5 \mod 11$

The answers are $x = \pm 6$ *and* $x = \pm 27$.

3.8 Exponentiation and Logarithm

Modular Exponentiation: The mathematical operation $y = a^x$ (where a is called base and x is called exponent) is called exponentiation. Exponentiation corresponds to the repeated multiplication of base to the exponent times i.e. $y = a^x = a \times a \times ... \times a (x$ times). In cryptography, we need to compute exponentiation under some modulo i.e. $y = a^x \mod n$. There is a very simple way to find modular exponentiation using "square and multiply" method.

3.8.1 Square and Multiply Method

Square and multiply method is an algorithm used to find large integer powers. It can quickly find powers even when the exponent has hundreds or thousands of digits, hence, the name "square and multiply" method.

Algorithm 13 Square and Multiply Method

Input- x, y, n; integers with $x > 0, y > 0, n > 1$;
Output- $x^y \mod n$;
mod $\_exp(x, y, n)$
1. If $y = 0$ return 1;
2. If $y \mod 2 = 0$ return $\mod \_exp(x^2(\mod n), y\%2, n); (y\%2 = \lfloor \frac{y}{2} \rfloor)$
3. Return $(x \mod \_exp(x^2 \mod n, y\%2, n)) \mod n$;

Example 62 *We want to find* 17^{23} mod 41. *Applying Algorithm 13 we get*

$$
\begin{aligned}
\text{mod } \_exp(17,23,41) &= 17. \quad \text{mod } \_exp(2,11,41); (17^2 \quad \text{mod } 41 = 2) \\
&= 17.2. \quad \text{mod } \_exp(4,5,41) \\
&= 17.2.4 \quad \text{mod } \_exp(16,2,41) \\
&= 17.2.4 \quad \text{mod } \_exp(38,1,41); (16^2 \quad \text{mod } 41 = 38) \\
&= 17.2.4.38 \quad \text{mod } 41 \\
&= 6
\end{aligned}
$$

By this example, we see that, if we find 17^{23} mod 41 directly then we need 22 multiplications modulo 41. But, using algorithm 13 we need to perform only 3 multiplications and 3 squarings under modulo 41.

Logarithm: The logarithm of a number is the exponent to another fixed number (called base). It is the inverse operation of the exponentiation operation i.e. for exponentiation $y = a^x$, the logarithm is $x = log_a y$. Logarithms were introduced by J. Napier in 17th century to simplify the calculations.

3.9 Discrete Logarithm Problem

In cryptography, factoring problem and discrete logarithm problem are two important number theoretic problems on which important cryptosystems are based. We have already discussed the factoring problem and algorithms for its solution in section 3.5. In this section, we will discuss the well-known discrete logarithm problem (DLP) and the related algorithms to solve it.

Definition 3.12   Discrete Logarithm Problem (DLP): Let G be a cyclic group generated by α. Then, for given $\beta \in G$ to find a positive integer a such that $\beta = \alpha^a$ in G, is called discrete logarithm problem (DLP).

It can be easily seen that $log_\alpha(\beta) = a$. Here, the underlying group structure is discrete, hence, the problem is called discrete logarithm problem . We can define DLP in a more general sense. In a group H, which is not necessarily cyclic, given $\alpha, \beta \in H$ to find a positive integer n such that $\beta = \alpha^n$ in H, is called DLP.

Example 63 *Consider* $G = Z_7^*$. *G itself is a cyclic group with generator 3. Since* $4 = 3^4$ mod 7, *thus* $log_3(4) = 4$ *in* Z_7^*.

DLP is one of the intractable problems in cryptography. Intractability of DLP depends upon the order of the group and exponent. Various theoretical approaches are suggested to solve DLP. In this section, we will discuss a few of them. To visualize DLP, readers may assume $G = Z_p^*$. Since exponentiation is a random function, α^i does not decrease monotonically with i, hence, we cannot use binary search to find i. In order to solve DLP, one has to perform an exhaustive search that might also include simple counting in the worst case $\alpha, \alpha^2, \alpha^3...$ until we get $\beta = \alpha^a$. If the order of the group is n, then using exhaustive search, the DLP can be solved in $\mathcal{O}(n)$ time (for n computation) and $\mathcal{O}(1)$ space (we need to store only the elements α, β, and α^n) assuming that multiplication of two elements requires a constant time i.e. $\mathcal{O}(1)$. Sometimes, this is called enumeration.

Example 64 *We determine the discrete logarithm of 171 to the base 5 in $Z^\star_{349}$. Enumeration yields $a = 300$ after computing 299 multiplication modulo 349.*

In cryptographic application, we take the size of exponent $a > 2^{160}$. Therefore, to solve DLP using enumeration, we need to compute at most $2^{160} - 1$ operations in the group. Thus enumeration is not feasible. Another approach is to pre-compute all possible values of α^i and store the ordered pair (i, α^i). After storing all the pairs, we sort the elements w.r.t. second parameter. Now we search for the given challenge β by employing binary search to find a such that $\alpha^a = \beta$. Using this algorithm, we need $\mathcal{O}(n)$ pre-computation time to compute n power of α and time $\mathcal{O}(n \log n)$ to sort the list of size n.

3.9.1 Shank's Baby-Step Giant-Step Algorithm

This is the first non-trivial algorithm to solve DLP. This is somewhat an improvement over the enumeration method. It requires fewer operations but more storage.

Algorithm 14 SHANKS ALGORITHM

INPUT-(G, n, α, β)
OUTPUT- $x = log_\alpha \beta$ i.e. x such that $\beta = \alpha^x$ in G
1. $m \leftarrow \lceil \sqrt{n} \rceil$
2. **for** $i \leftarrow 0$ to $m - 1$
do compute $\beta \alpha^{-i}$
3. Sort the m ordered pairs $(i, \beta \alpha^{-i})$ with respect to their second coordinate, obtaining a list L_1
4. **for** $j \leftarrow 0$ to $m - 1$
do compute α^{mj}
5. Sort the m ordered pairs (j, α^{mj}) with respect to their second coordinate, obtaining a list L_2
6. Find a pair $(j, y) \in L_2$ and a pair $(i, y) \in L_1$ (i.e. find two pairs having equal second coordinates)
7. $log_\alpha \beta \leftarrow (mj + i) \mod n$.

Explanation of Algorithm 14. Since $0 \leq log_\alpha \beta = x < n - 1$, if we divide x by $m = \lceil \sqrt{n} \rceil$, then we can express the unknown exponent $x = mj + i, 0 \leq i, j < m$. Since i is the remainder after division by m, clearly $0 \leq i < m$. Again, since $x \leq$

$n-1 \le m^2 - 1 = m(m-1) + m - 1$ implies that $0 \le j < m - 1$. In Algorithm 14 the steps 2 and 3 are called baby-steps and the steps 4 and 5 are called giant-steps. If we are able to find j and i, then we can compute the unknown exponent x. The baby steps and giant steps compute these values i and j. When we have computed i and j, then we have

$$\alpha^{mj+i} = \alpha^x = \beta \Rightarrow (\alpha^m)^j = \beta\alpha^{-i}$$

We first compute the baby-steps. If in the baby-steps list L_1 we find a pair $(i, 1)$, then we compute $\beta\alpha^{-i} = 1 \Rightarrow \beta = \alpha^i \Rightarrow x = i$. Thus, in the baby-steps we solve DLP if the unknown exponent $x < m$. If in the list L_1 there is no such pair, then in step 6, we search two pairs $(i, y) \in L_1$ and $(j, y) \in L_2$, then we have

$$\beta\alpha^{-i} = \alpha^{mj} \Rightarrow \beta = \alpha^{mj+i}.$$

As we have already explained that the unknown exponent x can be expressed as $x = jm + i$ with $0 \le i, j < m - 1$, the step 6 will be successful. However, if β is not an element of $< \alpha >$, then step 6 will not terminate successfully.

Example 65 *Consider $n = 31$ and let $\alpha = 3$ and $\beta = 6$. $m = \lceil\sqrt{31}\rceil = 6$. Then $\alpha^6 = 3^6 \mod 31 = 16$. First, we compute the ordered pair $(i, \beta\alpha^{-i})$ i.e. $(i, 6.3^{-1})$, for $0 \le i < 6$. We obtain the list $\{(0,6),(1,2),(2,11),(3,14),(4,15),(5,5)\}$. After sorting, we get list L_1. Now, we obtain the second list containing the ordered pair (j, α^{mj}) i.e. $(j, 3^{6j})$, for $0 \le j < 6$. Subsequently, we obtain the list $\{(0,1),(1,16),(2,8),(3,4),(4,2),(5,1)\}$. After sorting we get list L_2.*

Comparting list L_1 and L_2 we see that $(1,2) \in L_1$ and $(4,2) \in L_2$. Thus $i = 1$, and $j = 4$. Hence, $log_3 6 = x = 4 \times 6 + 1 = 25$ in Z_{31}^*. It can be easily verified that $3^{25} \equiv 6 \mod 31$.

We can compare this algorithm with enumeration. To compute the baby-steps, 6 multiplication modulo 31 is necessary. To compute the giant steps, 5 multiplication modulo 31 is necessary. However, for enumeration we require 25 multiplication modulo 31. Thus, enumeration requires more computation. Although, in Shank's algorithm we have to store 6 baby-step pairs, however, in enumeration we need to store only three elements.

Time and Space Complexity: If we observe Algorithm 14, we see that the steps 4 and 5 can be pre-computed. To store ordered pairs of baby-steps we require $\mathcal{O}(m)$ memory. The total time for step 4 is $\mathcal{O}(m)$. Step 2 also takes time $\mathcal{O}(m)$. Using best sorting algorithm, steps 3 and 5 can be done in time $\mathcal{O}(m \log m)$. Neglecting the $log\ m$ terms, we conclude that the algorithm runs in $\mathcal{O}(m)$ time with $\mathcal{O}(m)$ memory.

3.9.2 Pollard's Rho Algorithm

Explanation of Pollard's Rho Algorithm: The basic idea of this algorithm is to find a sequence $\{x_i, a_i, b_i\}_{i=0}$ such that $x_i = x_j, i < j$, and $x_i = \alpha^{a_i}\beta^{b_j}$. In order to save time, we seek a collision of the form $x_i = x_{2i}$. Let $(G, .)$ be a group and $\alpha \in G$ be

---

Algorithm 15 Pollard's Rho Algorithm for DLP

INPUT : (G, n, α, β); // (G be a group, α be a primitive root of order n, $\beta \in < \alpha >$)

OUTPUT : $log_\alpha \beta$;

Define the partition $G = S_1 \cup S_2 \cup S_3$ s.t. S_i and S_j are disjoint.

Define function $f(x, a, b)$

if $x \in S_1$

then $f \leftarrow (\beta x, a, b+1 \mod n)$;

else if $x \in S_2$

then $f \leftarrow (x^2, 2a \mod n, 2b \mod n)$;

else $f \leftarrow (\alpha x, (a+1) \mod n, b)$;

Return(f),

Main Steps

1. **initialize** $(x_0, a_0, b_0) \leftarrow (1, 0, 0)$, $i = 1$

2. $(x_i, a_i, b_i) \leftarrow f(x_{i-1}, a_{i-1}, b_{i-1})$

3. $(x_{2i}, a_{2i}, b_{2i}) \leftarrow f(f(x_{2(i-1)}, a_{2(i-1)}, b_{2(i-1)}))$

4. (a) **if** $\gcd(b_{2i} - b_i, n) \neq 1$;

return failure

(b) repeat the step 2 for $i = i + 1$

(c) **else return** $log_\alpha \beta = (a_i - a_{2i})(b_{2i} - b_i)^{-1} \mod n$

---

an element of order n. Let $\beta \in < \alpha >$ and we want to find $log_\alpha \beta$. Since $< \alpha >$ is a cyclic group of order n, we have $\log_\alpha \beta \in Z_n$. Assume that $c = \log_\alpha \beta$. To determine c, we divide G into three disjoint classes of equal size. Let $G = S_1 \cup S_2 \cup S_3$. Now, we define a function $f :< \alpha > \times Z_n \times Z_n \rightarrow < \alpha > \times Z_n \times Z_n$ as follows:

$$f(x, a, b) = \begin{cases} (\beta x, a, b+1) & \text{if } x \in S_1 \\ (x^2, 2a, 2b) & \text{if } x \in S_2 \\ (\alpha x, a+1, b) & \text{if } x \in S_3 \end{cases}$$

where each (x, a, b) satisfy the property $x = \alpha^a \beta^b$. Clearly,

If $x \in S_1 \Rightarrow x \rightarrow x\beta = \alpha^a \beta^{b+1} \Rightarrow a \rightarrow a \ \& \ b \rightarrow b+1$

If $x \in S_2 \Rightarrow x \rightarrow x^2 = \alpha^{2a} \beta^{2b} \Rightarrow a \rightarrow 2a \ \& \ b \rightarrow 2b$

If $x \in S_3 \Rightarrow x \rightarrow x\alpha = \alpha^{a+1} \beta^b \Rightarrow a \rightarrow a+1 \ \& \ b \rightarrow b$.

To construct the sequence $\{x_i, a_i, b_i\}$ having the desired property, we initialize the sequence by $(1, 0, 0)$. As per our definition of f, we observe that $f(x, a, , b)$ satisfies the desired property if (x, a, b) does. So we define,

$$(x_i, a_i, b_i) = \begin{cases} (1, 0, 0); & \text{if } i = 0 \\ f(x_{i-1}, a_{i-1}, b_{i-1}); & \text{if } i \geq 1 \end{cases}$$

As we have discussed in Pollard's Rho Algorithm for factoring, here also we compute the triples (x_{2i}, a_{2i}, b_{2i}) and (x_i, a_i, b_i) until we find a collision $x_{2i} = x_i$ for some $i \geq 1$. If we find a collision $x_{2i} = x_i$ for some i, then we have $\alpha^{a_{2i}} \beta^{b_{2i}} = \alpha^{a_i} \beta^{b_i} \Rightarrow \alpha^{a_{2i}+cb_{2i}} = \alpha^{a_i+cb_i} \Rightarrow c = (a_i - a_{2i})(b_{2i} - b_i)^{-1} \mod n$, if $\gcd(b_{2i} - b_i, n) = 1$.

Time and Space Complexity: Shank's Algorithm requires $\mathcal{O}(\sqrt{n})$ storage but Pollard's Rho Algorithm requires constant storage $\mathcal{O}(1)$. Time complexity of the Pollard's Rho Algorithm, to find a collision by birthday attack, is $\mathcal{O}(\sqrt{n})$, which is same as the complexity of Shanks's Algorithm.

Remark 3.11 If $G = Z_n$, then one of the simple ways to divide G into three disjoint classes is $S_1 = \{x : x \equiv 1 \mod 3\}$, $S_2 = \{x : x \equiv 0 \mod 3\}$ and $S_3 = \{x : x \equiv 2 \mod 3\}$.

Remark 3.12 If $1 \in S_2$ then $(x_i, a_i, b_i) = (1, 0, 0), \forall i \geq 1$.

Remark 3.13 Sometimes $\gcd(b_{2i} - b_i, n) \neq 1$ but the equation $(b_{2i} - b_i)c \equiv (a_i - a_{2i}) \mod n$ is solvable. In this case, we have to test each solution for the actual answer. See the following example.

Example 66 *Consider the cyclic group $G = Z_{1019}^\star$ which is generated by $\alpha = 2$. The order of the cyclic group is $n = 1018$. Let $\beta = 5$ be given. We have to find $\log_2 5$ in $Z_{1019}^\star$. Let us define three sets S_1, S_2 and S_3 as follows:*
$S_1 = \{x : x \equiv 2 \mod 3\}$,
$S_2 = \{x : x \equiv 0 \mod 3\}$,
$S_3 = \{x : x \equiv 1 \mod 3\}$.
After running the Algorithm 15 for $i = 1, 2, 3...$ we obtain triples (x_i, a_i, b_i) and (x_{2i}, a_{2i}, b_{2i}) as follows:

i	(x_i, a_i, b_i)	(x_{2i}, a_{2i}, b_{2i})
1	*(2,1,0)*	*(10, 1, 1)*
2	*(10, 1, 1)*	*(100, 2, 2)*
3	*(20, 2, 1)*	*(1000, 3, 3))*
4	*(100, 2, 2))*	*(425, 8, 6)*
5	*(200, 3, 2)*	*(436, 16, 24)*
.	...	...
.	...	...
.	...	...
50	*(505, 680, 378)*	*(101, 300, 415)*
51	*(1010, 681, 378)*	*(1010, 301, 416)*

The first collision in the above list is $x_{51} = x_{102} = 1010$. The equation to be solved is $(416 - -378)c = (681 - -301) \mod 1018 \Rightarrow 38c = 380 \mod 1018$. Since $\gcd(38, 1018) = 2 | 380$. The equation, reduced to $19c = 190 \mod 504 \Rightarrow c = $

$19^{-1}190 \mod 504 = 451 \times 190 \mod 504 = 10$. *We can easily check that* $2^{10} = 1024 \equiv 5 \mod 1019$. *Thus,* $log_2 5 = 10$ *is the answer. Another solution of the equation is* $c = 519$ *but* $2^{519} = -5 \mod 1019$, *which can be discarded.*

3.9.3 Pohlig-Hellman Algorithm

Now we will look at the Pohlig-Hellman Algorithm. This algorithm is a special - purpose algorithm for computing DLP in a multiplicative group whose order is a smooth integer. This algorithm was discovered by Roland Silver but first published independently by Stephen Pohlig and Martin Hellman. Hence, the algorithm is called Pohlig-Hellman algorithm. Let G be any cyclic group with generator α and of order n. Let $\beta = \alpha^a$ be given. To visualize the algorithm, we may assume G = $Z_p^\star$ and $n = p - 1$. By this algorithm, solving DLP in small order groups and using Chinese Remainder Theorem, we can solve the given DLP in a large order group.

Algorithm 16 Pohlig-Hellman Algorithm

INPUT - A generator α of a cyclic group G of order n and an element $\beta \in G$
OUTPUT - The discrete logarithm $a = \log_\alpha \beta$.

1. Find the prime factorization of $n = \prod_{i=1}^{k} p_i^{c_i}$

2. for $i = 1$ to k, do the following:

    (a) Compute $x_i = a_0 + a_1 p_i + a_2^2 p_i^2 + ... + a_{c_i-1} p_i^{c_i-1}$, where $x_i \equiv a \mod p_i^{c_i}$

    (b) (simplify the notation) Set $q \leftarrow p_i$ and $c \leftarrow c_i$;

    (c) $j \leftarrow 0, \beta_j \leftarrow \beta$ **while** $j \leq c - 1$

    do $\begin{cases} \delta = \beta_j^{\frac{n}{q^{j+1}}} \\ \text{find } i \text{ such that} \quad \delta = \alpha^{\frac{in}{q}} \\ a_j \leftarrow i \\ \beta_{j+1} \leftarrow \beta_j \alpha^{-a_j q^j} \\ j \leftarrow j+1 \end{cases}$

    Return $(a_0, a_1, a_2, ..., a_{c-1})$

    (d) Set $x_i = a_0 + a_1 q + a_2 q^2 + ... + a_{c-1} q^{c-1}$.

3. Using CRT compute the integer $x, 0 \leq x \leq n - 1$, such that $a \equiv x_i \mod p_i^{c_i}$ for $1 \leq i \leq k$.

4. Return(a).

Explanation of Algorithm 16. Pohlig-Hellman Algorithm solves discrete logarithm problem when n has only small prime factors. The basic idea of this algorithm is to iteratively compute the p-adic digits of the logarithm by repeatedly "shifting out" all but one unknown digit in the exponent, and computing that digit by elementary methods. It works for cyclic group G. If G is not cyclic, then we apply it to the subgroup $<g>$ of G generated by g which is always cyclic. Suppose $n = \prod_{i=1}^{k} p_i^{c_i}$. p_i's are distinct primes. Let p^c be one of the factors. Then, we compute $log_\alpha \beta$ under mod p^c. If this can be done for each $p_i^{c_i}$, then we can combine all these results by CRT and can get the required result.

Let us suppose that $n \equiv 0 \mod q^c$, $n \not\equiv 0 \mod q^{c+1}$. Here, first we will show how to compute $x \equiv a \mod q^c$, where $0 \leq x \leq q^c - 1$. Since, $x \leq q^c - 1$ we can express x in q-adic expansion as

$$x = \sum_{i=0}^{c-1} a_i q^i, \tag{3.1}$$

where the coefficients a_i is s.t. $0 \leq a_i \leq q - 1$ for $0 \leq i \leq c - 1$. Clearly, we can express a as $a = \sum_{i=0}^{c-1} a_i q^i + s q^c$ for some integer s. Thus, to compute x we need to compute $a_0, a_1, a_2, ... a_{c-1}$. The first step of the algorithm is to compute a_0. We claim that

$$\beta^{\frac{n}{q}} = \alpha^{a_0 \frac{n}{q}} \tag{3.2}$$

$$
\begin{aligned}
\beta^{\frac{n}{q}} &= (\alpha^a)^{\frac{n}{q}} \\
&= \left(\alpha^{a_0 + a_1 q + a_2 q^2 + ... + a_{c-1} q^{c-1} + s q^c}\right)^{\frac{n}{q}} \\
&= \alpha^{(a_0 + kq)\frac{n}{q}} (\text{for some integer} k) \\
&= \alpha^{a_0 \frac{n}{q}} . \alpha^{nk} \\
&= \alpha^{a_0 \frac{n}{q}}
\end{aligned}
$$

Using equation 3.2 we can determine a_0. This can be done by computing $\gamma = \alpha^{\frac{n}{q}}, \gamma^2, \gamma^3, ...$ until we get $\gamma^i = \beta^{\frac{n}{q}}$ for some $i \leq q - 1$. When this happens, we set $i = a_0$. In fact, this is a DLP in a cyclic group of small order q. If $c = 1$, we have done. For $c > 1$, we proceed to determine $a_1, a_2, ... a_{c-1}$. To determine these values we denote $\beta_0 = \beta$ and define-

$$\beta_j = \beta \alpha^{-(a_0 + a_1 q + a_2 q^2 + ... + a_{j-1} q^{j-1})} \tag{3.3}$$

for $0 \leq j \leq c - 1$. Now we generalize the equation 3.2 as follows:

$$\beta_j^{\frac{n}{q^{j+1}}} = \alpha^{a_j \frac{n}{q}} \tag{3.4}$$

It can be easily observed that equation 3.4 gets reduced to equation 3.2 when $j = 0$. Let us prove the equation 3.4. We have

$$
\begin{aligned}
\beta_j^{\frac{n}{q^{j+1}}} &= (\alpha^{a-(a_0+a_1q+a_2q^2+...+a_{j-1}q^{j-1})})^{\frac{n}{q^{j+1}}} \\
&= \alpha^{(a_jq^j+a_{j+1}q^{j+1}+...+s q^c)\frac{n}{q^{j+1}}} \\
&= \alpha^{(a_jq^j+k_jq^{j+1})\frac{n}{q^{j+1}}} \text{ (for some integer } k_j) \\
&= \alpha^{a_j\frac{n}{q}}
\end{aligned}
$$

Hence, for a given β_j, we can compute a_j from equation 3.4. Also using equation 3.5 given β_j we can compute β_{j+1} if a_j is known.

$$\beta_{j+1} = \beta_j \alpha^{-a_jq^j} \tag{3.5}$$

Therefore, we can compute $a_0, \beta_1, a_1, \beta_2, a_2, \beta_3, a_3, ... \beta_{c-1}, a_{c-1}$ by applying equation 3.4 and 3.5.

Example 67 *Suppose $p = 29$ is a prime and $\alpha = 2$ is a primitive root mod p. We consider $\beta = 18$. Then $x = log_2 18$ is computed as follows-*
Step-1. *The prime factorization of $n = p - 1 = 28 = 2^2.7^1$. Clearly, $q_1 = 2, q_2 = 7$ and $c_1 = 2, c_2 = 1$.*
Step-2.1. *We first compute $x = x_1 \mod 2^2$ for $q = 2, c = 2$. By applying algorithm, we find that $a_0 = 1, a_1 = 1$ and so $x \equiv 3 \mod 4$ [since, $18^{14} \mod 29 = 28$, $2^{14} \mod 29 = 28 \Rightarrow a_0 = 1$ and $\beta_1 = 18.2^{-1} = 9$ and $9^7 \mod 29 = 28$, $2^{14} \mod 29 = 28 \Rightarrow a_1 = 1$].*
Step-2.2. *We then compute $x \equiv x_2 \mod 7$ for $q = 7, c = 1$. Applying the above algorithm, we find that $a_0 = 4$. Hence, $x \equiv 4 \mod 7$.*
Step-3. *Finally, we will the system of equations $x \equiv 3 \mod 4$ and $x \equiv 4 \mod 7$ using CRT and get $x \equiv 11 \mod 28$. Thus, $log_2 18 = 11$.*

3.9.4 Index Calculus Algorithm

This is a probabilistic and specialized method in $Z_p^\star$ (p is prime) for solving discrete logarithmic problem. This algorithm collects relations among discrete logarithms of small primes, computes them by a linear algebra method, and finally expresses the desired discrete logarithm with respect to the discrete logarithms of small primes. Let α be a primitive root of $Z_p^\star$ and $\beta = \alpha^a \mod p$. Our aim is to find a. Now, to compute a, we follow the procedure below:

1. First we consider the factor base $\mathcal{B} = \{p_1, p_2, ..., p_k\}$.

2. Then, compute $log_\alpha p_i$ for $i = 1, 2, ..., k$.

3. Using $log_\alpha p_i$, compute $log_\alpha \beta$.

The steps 1 and 2 are pre-computable steps.

Let us see to compute $log_\alpha p_i$, $i = 1,2,3,...k$. For this, we construct c congruence modulo p equations with $c > k$ with variables $log_\alpha p_i$. In general, we take $c = 4k$. The c congruence equations modulo p are in the form $\alpha^{x_j} \equiv p_1^{a_{1j}} p_2^{a_{2j}} ... p_k^{a_{kj}}$ mod p, where $1 \leq j \leq c$, which implies that $x_j \equiv a_{1j} log_\alpha p_1 + a_{2j} log_\alpha p_2 + ... + a_{kj} log_\alpha p_k$ mod $(p-1)$. Thus, we have c congruences in k unknown $log_\alpha p_i$ $(1 \leq i \leq k)$. Solving these system of equations, we get a unique solution mod $(p-1)$.

Now, the question is how to generate c congruences of the desired form? One of the ways is to take a random value x and compute α^x and then determine if α^x mod p has all its factors in B. Suppose, we have successfully carried out the pre-computation steps 1 and 2. Finally, we compute the desired algorithm $log_\alpha \beta$ using a Las-Vegas (please refer definition 5.22) type algorithm. The steps are as below.

Step-3.1. Choose a random integer $s(1 \leq s \leq p-2)$ and compute $\gamma = \beta \alpha^s$ mod p.

Now, attempt to factor γ over the factor base if this can be done, then $\gamma = p_1^{c_1} p_2^{c_2} .. p_k^{c_k}$, which implies that $\beta \alpha^s \equiv p_1^{c_1} p_2^{c_2} .. p_k^{c_k}$ mod p, otherwise repeat the process for different values of s.

Step-3.2. Compute $s + log_\alpha \beta = c_1 log_\alpha p_1 + ... + c_k log_\alpha p_k$ mod $(p-1)$. Which implies that $log_\alpha \beta = [(c_1 log_\alpha p_1 + ... + c_k log_\alpha p_k) - s]$ mod $(p-1)$. Let us do the following example.

Example 68 *Suppose we take $p = 29$ which is a prime. A primitive root is $\alpha = 2$. Let $\beta = 18$ be given. The question is to find $log_2 18$. Let us take the factor base $B = \{2,3\}$. We are using lucky exponent 3 and 5. We have*
$\alpha^3 = 2^3 \equiv 8$ mod $29 = 2^3$ *and*
$\alpha^5 = 2^5 \equiv 3$ mod $29 = 3$

From the above two equations, we have $log_2 2 = 1$ and $log_2 3 = 5$. Now, with choose $s = 4$, we have $\beta \alpha^s$ mod $p = 18 \times 2^4$ mod $29 = 27 = 3^3$. On solving, we get $4 + log_2 18 = 3 log_2 3$ mod $28 \Rightarrow log_2 18 = 11$.

3.10 Exercises

1. Compute the gcd of the following pairs using Euclidean Algorithm:
   (a) 44 and 320 (b) 561 and 220(c) 400 and 564 (d) 700 and 300.

2. Use CRT to find the number which gives the remainder 3 when divided by 7, 6 when divided by 10, and 3 when divided by 13.

3. Using Extended Euclidean Algorithm, express the gcd of the following pairs as their linear combination:
   (a) 81 and 27(b) 24 and 320.

4. Determine for which of the following equations the value of x exists, if exist

find all such x and if not then explain why.

(a) $5x \equiv 12 \mod 21$ (b) $64x \equiv 25 \mod 21$ (c) $15x \equiv 12 \mod 57$.

5. In $\mathbb{Z}_{13}^*$ find the numbers which are quadratic residue and which are not quadratic residue.

6. Using quadratic residue , solve the following

   (a) $x^2 \equiv 12 \mod 34$ (b) $x^2 \equiv 7 \mod 13$

7. Compute using Fermet's little theorem

   (a) $145^{18} \mod 17$ (b) $15^{15} \mod 13$.

8. Find $z = x + y$, where $x = 132$ and $y = 270$ but the system can take input only less than 100. (Hint: Use CRT)

9. Using Square and Multiply Algorithm compute

   (a) $21^{23} \mod 8$ (b) $2007^{41} \mod 2000$.

10. Find x in following cases

    (a) $2 \equiv 5^x \mod 7$ (b) $3 \equiv 7^x \mod 10$.

11. Use Pollard's Rho Algorithm to find the factor of

    (a) 3731 (b) 434617.

12. Compute the following

    (a) $16^{-1} \mod 101$ (b) $36^{-1} \mod 77$

13. Which of the following satisfy the Miller-Rabin test with base 2

    (a) 349 (b) 109 (c) 561 .

14. Which of the following satisfy the Fermet's primality test

    (a) 271 (b) 341 (c) 561.

15. Compute $7^{120007} \mod 143$ using Chinese Remainder Theorem and Fermat's Little Theorem.

Chapter 4

Probability and Perfect Secrecy

CONTENTS

The cryptosystems those have been discussed in Chapter 6 are not perfectly secure. They can be broken using ciphertext only attack or known plaintext attack etc. The concept of perfect secrecy was proposed by Claude Shannon in 1949 [155]. A cryptosystem is said to be perfectly secure if the attacker cannot get any information about the plaintext by observing only the ciphertext. We can understand the notion of perfect secrecy mathematically by the concept of probability theory. The probability theory is applicable in cryptography usually to estimate "how probable it is that an insecure event may occur under certain conditions". It is also applicable to analyze some security notions such as *unconditional security*, *semantic security* etc. of any cryptosystem.

4.1 Basic Concept of Probability

Probable means "likely" or "most likely" to be true though not certain to occur. Measure of "chance" or "likelihood" for the statement to be true is called probability of the statement. A probability model consists of three components: sample space, event, and a probability function or experiment. Sample space S is a finite non-empty set. Elements of this set are called outcomes or points or elementary events. The set E of events is a subset of the sample space S. Probability function P (also called Experiment) is a function that assigns probabilities (real numbers between 0 and 1) to the events in E.

Definition 4.1   Experiment: Experiment means an approach which can produce some well-defined outcomes.

There are basically two types of experiments; Deterministic experiment and Probabilistic or Random experiment. Any experiment or event under uncertain situations is called a random trial or random experiment, otherwise the experiment is called a deterministic experiment. In probability theory, we call the experiment as a random experiment.

Definition 4.2   Elementary event: If a random experiment is performed, then each of its outcomes is known as an elementary event.

Definition 4.3   Sample space: The set of all possible outcomes of a random experiment is called a sample space associated with it. It is denoted by S.

Definition 4.4   Event: A subset of a sample space associated with a random experiment is called an event.

An event associated to a random experiment is said to occur if any one of the elementary events associated to it is an outcome.

Example 69 *If we flip a coin, flipping coin is an experiment, then the sample space is S: {head, tail}. One of the events, getting head is E_1: {head}.*

Example 70 *If we throw a dice, then throwing dice is an experiment, and the sample space is S : {1,2,3,4,5,6}. One of the events getting at least 3 is E_2 : {3,4,5,6}.*

Definition 4.5   Probability: Suppose, an experiment can yield n equally probable points and every experiment must yield a point, then the sample space is $S = (a_1, a_2, ...a_n)$. Let m be the number of points which form an event E. Then *probability of occurring of an event E* is denoted by $P(E)$ and given by $P(E) = \frac{n}{m}$ i.e. ratio of favourable elementary events. The total possible elementary events is called probability of the given event.

In the above examples 69 and 70, the probability of events are $P(E_1) = \frac{1}{2}$ and $P(E_2) = \frac{4}{6}$.

Definition 4.6   Conditional probability: If A and B are two events and $P(B) > 0$, then the conditional probability of A, given B is defined as follows: $P(A|B) = P(A \cap B)/P(B)$.

Two events A and B are said to be independent if $P(A \cap B) = P(A)P(B)$. Thus, if the events A and B are independent then $P(A|B) = P(A)$.

Theorem 4.1
The Law of Total Probability. *If* $S = \cup_{i=1}^{n} E_i$ *and* $E_i \cap E_j = \phi$ $(i \neq j)$, *then for any event* $A \subset S$, $P(A) = \sum_{i=1}^{n} P(A|E_i)P(E_i)$.

Proof 4.1 For any $A \subset S$, we have $A = S \cap A = \cup_{i=1}^{n} S \cap E_i$, which implies that $P(A) = \sum_{i=1}^{n} P(A \cap E_i) = \sum_{i=1}^{n} P(A|E_i)P(E_i)$.

Definition 4.7   Random Variable: The variable which assigns a real value in the random experiment is called a random variable.

The cryptographic functions are mainly defined on discrete structure/space such as a finite set of integers between an interval, finite field, and finite group. Below, we describe the variables defined on discrete space.

Definition 4.8   Discrete Random Variable: A discrete random variable is a numerical result of an experiment defined on a discrete sample space. In other words, a discrete random variable, say X, consists of a finite set S and a probability distribution defined on S. The probability that the random variable X takes the value x_i is denoted by $P[X = x_i]$.

Definition 4.9   Probability Distribution Function: A probability distribution function f defined on the sample space S is a real valued function satisfying the conditions : (1) $0 < P[X = x_i]$, (2) $\sum_x P[X = x] = 1$.

Remark 4.1   Let the random variable X be defined on S and $E \subset S$. The probability that X takes a value in the subset E is computed by $P[E] = \sum_{x \in E} P[X = x]$. The subset E is called an event.

Suppose X and Y are two random variables defined on finite sets S and T respectively. In terms of random variable, we can denote the joint and conditional probability as follows:

Joint Probability: The joint probability that the random variable X takes the value x and the random variable Y takes the value y is denoted by $P[x,y]$.

Conditional Probability: The probability that X takes the value x given that Y takes the value y is denoted by $P[x|y]$.

Independent Random Variables: Two random variables X and Y are said to be independent random variables if $P[x,y] = P[x]P[y]$, where $x \in S$ and $y \in T$.

Relation between Joint and Conditional Probability: $P[x,y] = P[x|y]P[y] = P[y|x]P[x]$.

Definition 4.10   Uniform Distribution: Let S be any sample space having n distinct elements and X be a random variable defined on S. If $P[X = x_i] = \frac{1}{n}, \forall i$, then we say that the random variable X is uniformly distributed on the sample space S. We denote $x \in_u S$, if the point x is chosen randomly from a uniformly distributed sample space S.

4.2   Birthday Paradox

Birthday paradox is one of the good examples to visualize the application of probability theory in cryptography. It is also useful to check security of a cryptosystem under collision attack also called birthday attack. Consider a function $f : X \rightarrow Y$, where Y is a set of n distinct elements. Birthday paradox is: "How many distinct values we have to choose uniformly from X so that we get a collision of f with certain probability bound" i.e. for a given probability bound ε (i.e. $0 < \varepsilon < 1$), to find a value k such that for k pairwise distinct values $x_1, x_2, ...x_k \in_u X$, the k evaluations $f(x_1)$, $f(x_2)$,...$f(x_k)$ satisfy $P[f(x_i) = f(x_j)] \geq \varepsilon$ for some $i \neq j$. To enlarge the value of k the function f must be a randomized function. Also, to find a collision, it is necessary that $|X| > |Y|$. If we consider X as a group of people, the image $f(x_i)$ as the birthday of ith person x_i and Y as a set of n different birthdays, then the birthday paradox becomes "what is the minimum number of people necessary in a group for the fact that probability that two people have same birthday is greater than some fixed ε"? To answer this question, we construct elementary events and probability space. Suppose, we choose k values of x_i from X. Then an elementary event is $(y_1, y_2, y_3...y_k) \in \{1,2,3,...n\}^k$, where $y_i = f(x_i)$. In this way, we have total n^k elementary events. Assuming that all elementary events are equally probable, the probability of an elementary event is $\frac{1}{n^k}$. We want to compute the probability of $y_i = y_j, i \neq j$. Let $P[y_i = y_j] = p$, then $q = 1 - p$ is the probability $P[y_i \neq y_j], i \neq j$. Our interest is to estimate the probability $q < 1 - \varepsilon$, so that $p \geq \varepsilon$. Therefore, we consider $E = (y_1, y_2, y_2....y_k) \in \{1,2,3...n\}^k; y_i \neq y_j, 1 \leq i, j \leq k$. For any element $(y_1, y_2, y_2....y_k)$ in E, there are n possibilities for the first entry, $n - 1$ possibilities for the second entry, and so on. Thus for the ith entry of any element in E, there are $n - (i - 1)$ possibilities. Hence, we obtain

$$|E| = \prod_{i=1}^{k-1}(n-i). \tag{4.1}$$

and

$$q = \frac{1}{n^k}\prod_{i=1}^{k-1}(n-i) \tag{4.2}$$

$$= \prod_{i=1}^{k-1}\left(1-\frac{i}{n}\right) \tag{4.3}$$

$$\le \prod_{i=1}^{k-1}e^{\frac{-i}{n}} \tag{4.4}$$

$$= e^{-\sum_{i=1}^{k-1}\frac{i}{n}} \tag{4.5}$$

$$= e^{-\frac{k(k-1)}{2n}} \tag{4.6}$$

Thus

$$q < e^{-\frac{k(k-1)}{2n}} \tag{4.7}$$

We want to determine k such that $q < (1-\varepsilon)$. By equation 4.7 it is sufficient to consider $e^{-\frac{k(k-1)}{2n}} < 1-\varepsilon$ i.e. $-\frac{k(k-1)}{2n} < log(1-\varepsilon)$, or $k(k-1) \ge 2nlog\frac{1}{1-\varepsilon}$. Which can be obtained if we choose

$$k \ge \frac{1+\sqrt{1+8nlog\frac{1}{1-\varepsilon}}}{2} \tag{4.8}$$

For $\varepsilon = 1/2$, we need to choose $k \ge \frac{1+\sqrt{1+8nlog\,2}}{2}$ i.e. if we choose approximately $\sqrt{n}$ number of elements from X, then there is more than 50% chance that we get a collision $f(x_i) = f(x_j)$ for some $(x_i \ne x_j)$.

Remark 4.2 For $n = 365$ and $\varepsilon = 1/2$, we get $k \ge 23$. Thus, if we choose at least 23 people in a room then there is more than 50% chance that two people will have their birthday on the same date. Suppose the problem is to find x for given $f(x)$, where $x \in [a,b]$. Clearly, evaluation of $f(a), f(a+1), f(a+2), \dots$ can reveal x before $b-a$ steps. If $b-a$ is too large, then this exhaustive search cannot be practical. However, if $\sqrt{b-a}$ is a tractable value (e.g. $\sqrt{b-a} = 2^{50}$), then birthday paradox can play a role in inverting $f(x)$ in 2^{50} steps. If f is an encryption function, then to prevent it from birthday attack the size of the ciphertext space $|C|$ should be sufficiently large (e.g. $|C| > 2^{160}$).

4.3 Perfect Secrecy

In this section, we will discuss the concept of perfect secrecy proposed by Shannon. Consider a cryptosystem $(\mathcal{P},\mathcal{C},\mathcal{K},\mathcal{E},\mathcal{D})$, where $\mathcal{P}$ is the plaintext space, $\mathcal{C}$ is the ciphertext space, $\mathcal{K}$ represents the key space, $\mathcal{E}$ is the set of all encryption functions and $\mathcal{D}$ is a finite set of all decryption functions. Since plaintext, key, and ciphertext character are chosen from some fixed alphabet having certain probability distribution e.g. English letters having frequency distribution, we define random variables X, K and Y on plaintext space, key space, and ciphertext space independently. We denote the priori probability that plaintext x occurs by $P[X = x]$, the priori probability that a key k is chosen by $P[K = k]$ and the priori probability that a ciphertext y is chosen by $P[Y = y]$. To encrypt any plaintext x, we choose a key k from key space then using encryption function e_k we get a ciphertext $y = e_k(x)$ in the ciphertext space $\mathcal{C}$. So, the probability distribution on plaintext and key space induces a probability distribution on ciphertext.

For any $k \in \mathcal{K}$, the collection of all possible ciphertexts using key k is defined by $C(k) = \{e_k(x) : x \in \mathcal{P}\}$.

Since a ciphertext y is obtained by a plaintext x and a key k, we have for every $y \in C$.

$$P[Y = y] = \sum_{\{k:y\in C(k)\}} P[K = k]P[x = d_k(y)] \tag{4.9}$$

Thus, the probability distribution of ciphertext space depends on the probability distribution of plaintext and key space. The question is whether the ciphertext leaks any information about the plaintext? To answer this question, we find the conditional probability of plaintext $X = x$, given the ciphertext $Y = y$. This probability is called a posteriori probability of the plaintext. We compare the posteriori probability of plaintext with the priori probability of plaintext. If both the probabilities are equal, then we conclude that the random variables X and Y are independent, and the ciphertext does not modify the probability distribution of plaintext space. For this, we first compute the conditional probability of ciphertext $Y = y$, given the plaintext $X = x$ as follows:

$$P[Y = y|X = x] = \sum_{\{k:x=d_k(y)\}} P[K = k] \tag{4.10}$$

Thus, by Bay's theorem we obtain

$$P[X = x|Y = y] = \frac{P[X = x]P[Y = y|X = x]}{P[Y = y]} \tag{4.11}$$

Let us look at an example to understand the probability distribution.

Example 71 *Let* $\mathcal{P} = \{a,b,c\}$ *with* $P[a] = 1/4$, $P[b] = 1/2$, $P[c] = 1/4$. *Let* $\mathcal{K} = \{k_1, k_2, k_3\}$ *and* $\mathcal{C} = \{1,2,3,4\}$ *and* $P[k_1] = 1/2$, $P[k_2] = P[k_3] = 1/4$. *The cryptosystem is defined as*

	a	b	c
k_1	1	2	3
k_2	2	3	4
k_3	3	4	1

Let us find the probability distribution of ciphertext space $\mathcal{C}$ *as* $P[1] = \frac{3}{16}, P[2] = \frac{5}{16}, P[3] = \frac{5}{16}$ *and* $P[4] = \frac{3}{16}$.

The conditional probability distribution of plaintext that a certain ciphertext has been observed is given by:
$P[a|1] = \frac{2}{3}, P[a|2] = \frac{1}{5}, P[a|3] = \frac{1}{5}, P[a|4] = 0,$
$P[b|1] = 0, P[b|2] = \frac{4}{5}, P[b|3] = \frac{2}{5}, P[b|4] = \frac{2}{3},$
$P[c|1] = \frac{1}{3}, P[c|2] = 0, P[c|3] = \frac{2}{5}, P[c|4] = \frac{1}{3}.$

Now, we are all set to understand "Perfect Secrecy". Informally, perfect secrecy refers to the fact that an attacker cannot obtain any information about the plaintext by observing the ciphertext. Using the concept of probability distribution, we can define the perfect secrecy as follows:

Definition 4.11 **Perfect Secrecy:** A cryptosystem has perfect secrecy if $P[x|y] = P[x]$, for all $x \in \mathcal{P}$ and $y \in \mathcal{C}$. That is, a posteriori probability that the plaintext is x, given that the ciphertext y is observed, is identical to the priori probability that the plaintext is x.

In the above Example 71, none of the ciphertexts satisfies perfect secrecy.

Remark 4.3 By the Bayes' theorem we observe that if $P[x|y] = P[x]$, $\forall x \in \mathcal{P}$, and $y \in \mathcal{C}$, then $P[y|x] = P[y]$, $\forall x \in \mathcal{P}$, and $y \in \mathcal{C}$ also.

Remark 4.4 Since $P[y] > 0$, for each $y \in \mathcal{C}$, there must be at least one $k \in \mathcal{K}$ such that $e_k(x) = y$. This implies $|\mathcal{K}| > |\mathcal{C}|$. Again, since encryption rule is injective, we have $|\mathcal{C}| > |\mathcal{P}|$. Thus, $|\mathcal{K}| > |\mathcal{C}| > |\mathcal{P}|$. If $|\mathcal{K}| = |\mathcal{C}| = |\mathcal{P}|$, then we get the following theorem due to Shannon.

Theorem 4.2
A cryptosystem $(\mathcal{P}, \mathcal{C}, \mathcal{K}, \mathcal{E}, \mathcal{D})$, *where* $|\mathcal{K}| = |\mathcal{C}| = |\mathcal{P}|$, *provides perfect secracy if and only if every key is used with equal probability* $\frac{1}{|\mathcal{K}|}$ *and for every* $x \in \mathcal{P}$ *and* $y \in \mathcal{C}$, *there is a unique key* $k \in \mathcal{K}$ *such that* $e_k(x) = y$.

Proof 4.2 Suppose, the given cryptosystem provides perfect secrecy. Consider a pair of plaintext and ciphertext (x,y). By the encryption rule, for each $x \in \mathcal{P}$ and $y \in \mathcal{C}$; there must be at least one key k such that $e_k(x) = y$. According, to our assumption $|\mathcal{C}| = |\mathcal{K}|$ i.e. $|\{e_k(x) : k \in \mathcal{K}\}| = |\mathcal{K}|$, we conclude that there is exactly one key k such that $e_k(x) = y$. Now let us denote $n = |\mathcal{K}|$ and let $P = \{x_i : 1 \leq i \leq n\}$. Fix a ciphertext element $y \in \mathcal{C}$. Since, each pair (x_i, y) of plaintext and ciphertext corresponds to a unique key in the key space, we can arrange the keys in the key space as $k_1, k_2, \ldots k_n$ such that $e_{k_i}(x_i) = y, 1 \leq i \leq n$. Since the system is perfectly secure, by Baye's theorem we have

$$p[x_i] = P[x_i|y] = \frac{P[x_i]P[y|x_i]}{P[y]} = \frac{P[x_i]P[k = k_i]}{P[y]} \tag{4.12}$$

From above equation 4.12, we get $P[k_i] = P[y]$, $\forall i, 1 \leq i \leq n$. Hence, we conclude that the keys are uniformly distributed over the key space with $P[K = k_i] = \frac{1}{|\mathcal{K}|}$. This proves the first part of the theorem.

Conversely, assume that the keys are uniformly distributed over the key space with $P[K = k] = \frac{1}{|\mathcal{K}|}$ and for every pair of plaintext and ciphertext (x,y), there is a unique key k such that $e_k(x) = y$. Then we have to prove that the system is perfectly secure i.e. $P[x|y] = P[x]$ for any plaintext $x \in \mathcal{P}$ and ciphertext $y \in \mathcal{C}$. For any ciphertext $y \in \mathcal{C}$, we get from equation 4.9 that $P[Y = y] = P[K = k : y = e_k(x)]$. By Baye's theorem we have

$$P[x|y] = \frac{P[x]P[y|x]}{P[y]} = \frac{P[x]P[k : y = e_k(x)]}{P[y]} = P[x].$$

This completes the proof.

Remark 4.5   Using this theorem we can prove that a shift cipher is perfectly secure if all 26 keys are used with equal probability $\frac{1}{26}$.

4.4   Vernam One-Time Pad

Vernam one-time pad was invented by Gilbert Vernam in 1917. In this cryptosystem, $\mathcal{P} = \mathcal{C} = \mathcal{K} = \{0,1\}^n$. Thus, in Vernam one-time pad we encrypt a bitstrings of length n using a key of n bitstrings and get a ciphertext of n bitstrings. Let $x = (x_1, x_2, x_3 \ldots x_n) \in \{0,1\}^n$ be any plaintext. Before encrypting x, the sender first chooses a share of secret key $k = (k_1, k_2, k_3 \ldots k_n) \in_U \{0,1\}^n$. The encryption and decryption is performed as follows.

- **Encryption**–$e_k(x) = x \oplus k$, where $\oplus$ operation is performed component-wise.

- **Decryption**–$d_k(y) = y \oplus k$, where $\oplus$ operation is performed component-wise.

We can choose n-bits from $\{0,1\}^n$ in 2^n different ways. Hence, the shared secret keys are uniformly distributed over the key space $\{0,1\}^n$ with equal probability $\frac{1}{2^n}$.

By using the the Shannon's theorem 4.2 for perfect secrecy, the Vernam one-time pad is perfectly secure. When Vernam introduced his cryptosystem, he conjectured that the cryptosystem is perfectly secure and no one can break his cryptosystem. But, he failed to prove his conjecture. After 40 years of Vernam one-time pad, in 1949 Shannon proved that the Vernam one-time pad has perfect secrecy.

One of the drawbacks of this cryptosystem is that the encryption key can be used only once. After one encryption, the shared secret key must be discarded. Hence, the cryptosystem is called "one-time pad". Due to this, the crytosystem is impractical. Just one pair of plaintext and ciphertext is sufficient to obtain the shared secret key as $x \oplus y = k$. Thus known plaintext attack is possible in this case. Hence, the repeatation of secret key is not allowed.

4.5 Random Number Generation

In the construction of one-time pad and many other cryptographic applications (key generation in AES, DES etc.), we need to generate a random number of a given size. For this, we first generate a random bitstring of size n. Let the random bitstring is $\{b_1, b_2, b_3...b_n\}$, then we compute $a = \sum_{i=1}^{n} b_i 2^i$. Clearly, $a \in \{0, 1, 2, ...m\}$ where $n = \lfloor m \rfloor + 1$. In this way, we can generate a uniformly distributed random number of size n. How to generate a random bitstring? Coin flipping is one of the good examples where a random bitstring is generated. We flip a coin and see whether it is a head or tail. For head, we write "1" and for tail we write "0". In this way, if we flip a coin n times then we can generate a truly random bitstring of size n. But flipping a coin is not practical. In general, we need to generate bistring of large size using some secure software/hardware so that no one can predict the output bits. It is time consuming to generate a true random number. So, we use a pseudo-random number generator.

4.6 Pseudo-random Number Generator

A pseudo-random number generator is an algorithm that produces a long sequence of bits from a given short sequence of random bits (called initial seed). The output sequence of bits is not truly random but looks like a random sequence (called pseudo-random bitstring) as it depends on an initial seed. The algorithm is called pseudo-random number generator . A non secure example of pseudo-random number generator can be obtained as follows. Suppose, we want to generate a pseudo-random number $a < m$. We choose $x_0 < m$ as an initial seed. Define $x_n = ax_{n-1} + b(mod\ m)$, where a and b are given integers less than m. We run the algorithm for $n = 1, 2, 3...m$. The output x_n is a pseudo-random number less than m. It has been shown that any polynomial congruential generator in cryptography is insecure, hence, the output bits are predictable with high probability. To generate a cryptographically secure pseudo-random number, either we use a one-way function or an intractable problem from number theory. The well-known and popular Blum-Blum-Shub (BBS) [23] pseudo-random bit generator is based on quadratic residue problem. In this technique, we first generate two large primes p and q such that $p \equiv q \equiv 3 (mod\ 4)$. Then, we set $n = pq$ (the integer n of this type is called Blum integer). Then, we choose a random initial seed $x_0 \in_U Z_n$. Further, we define recursively as $x_i \equiv x_{i-1}^2 (mod\ n)$ and $b_i \equiv x_i (mod\ 2)$. The pseudo-random bits are $\{b_1, b_2, b_3....\}$. It has been proved that the BBS pseudo-random number generator is unpredictable [23]. Let us do the following example.

Example 72 *Choose primes $p = 83$ and $q = 43$ and set $n = p \times q = 3569$. Let the initial seed be $x_0 = 5$. Using the recurrence relation $x_i \equiv x_{i-1}^2 (mod\ n)$, we obtain the following sequence of x_is.*
$x_1 = 25, x_2 = 625, x_3 = 1604, x_5 = 1901, x_6 = 1973, x_7 = 2519, x_8 = 3248, x_9 = 3109, x_{10} = 1029...., thus, the pseudo-random bitstring generated is 110111011....$

4.7 Exercises

1. Let S be family with two children = $\{gg, gb, bg, bb\}$ where g- girl, b- boy in $ab-a$ is elder than b. If A = family has a girl and B = both children are girls. Determinie $P(B|A)$.

2. Let $p = 2q + 1$, such that both p and q are prime numbers. Consider choosing two numbers g and h from the set $S = \{1, 2, 3..., p-1\}$ with replacement. Let event A be "h is generated by g", that is $h = g^x mod\ p$ for some $x < p$. What is the probability of A for random g and h.

3. Let S be the set of non-negative numbers up to k bits (binary digits). Sample a point in S at random by following the uniform distribution. Show that the probability of the sample point is k-bit number is 1/2.

4. Let a fair coin be tossed 20 times. What is the probability for all possible number of head appearance.

5. Using coin flipping, generate two random numbers of size 50 bits.

6. Using Blum-Blum-Shud random numbers generation algorithm, generate a random number of 100 bits.

7. Model an algorithm using birthday attack to solve discrete logarithm problem. (given h, g and p, find x such that $h \equiv g^x\ mod\ p$).

8. Determine k such that the probability of two of k people having same birthday is at least $\frac{8}{10}$.

9. Consider a cryptosystem, with plaintext $\mathcal{P} = \{a, b\}$, ciphertext $\mathcal{C} = \{1, 2, 3, 4\}$ and key space $K = \{k_1, k_2, k_3\}$. The probability distribution in plaintext space is $P[a] = \frac{3}{4}$, $P[b] = \frac{1}{4}$. The probability distribution in key space is $P[k_1] = P[k_2] = \frac{1}{4}, P[k_3] = \frac{1}{2}$. The cryptosystem is defined as

	a	b
k_1	1	2
k_2	3	4
k_3	2	3

Find the probability distribution of ciphertext space and the conditional probability distribution of plaintext for a given ciphertext.

10. Show that the Shift cipher in Z_{26} is perfectly secure if each key is equally distributed with equal probability over the key space.

11. Why Caesar cipher is not perfectly secure?

12. Suppose that eight digit alphanumeric passwords are randomly distributed among people. What should be the minimum number of people in a room so that the probability of two people have same password is at least $\frac{1}{2}$?

13. Consider the following variation of the one-time pad cryptosystem. Let $\mathcal{P} = \mathcal{K} = \{00, 01, 10\}^l$, $l > 0$. Encryption and decryption works in the same way as in the one-time pad. Decide whether this cipher is perfectly secure or not. Explain your answer.

Chapter 5

Complexity Theory

CONTENTS

The word *complexity* in the theory of computation is probably due to the inherent difficulty in a computational problem. This theory is important to study the behavior of the problems and to quantify the amount of resources such as time and storage needed to solve them. When there are several algorithms for the same objective then for various reasons such as efficiency we need to compare the running time of different algorithms. Theoretically, we compare the algorithms by computing time complexity. We refer the term *computational complexity* of an algorithm to be the total number of steps required for the execution of the algorithm. Below, we define various terms more precisely.

Definition 5.1   An algorithm is a step-by-step procedure for solving a problem in a finite amount of time.

5.1 Running Time and Size of Input

Running time of any algorithm depends on the size of the input of the algorithm.

Definition 5.2   Size of Input: The 'size of input' is the total number of bits needed to represent the input in ordinary binary notation using an appropriate encoding scheme.

We will use *log* for log_2 and *lg* for log_e throughout the chapter. See the following examples.

Example 73 *The number of bits in the binary representation of a positive integer n is $1 + \lfloor logn \rfloor$ bits. The size of number n is approximated by log n.*

Example 74 *If f is a polynomial of degree at most k, each coefficient being a nonnegative integer at most n, then the size of f is $(k + 1)log\ n$ bits.*

Example 75 *If A is a matrix with r rows, c columns, and with non-negative integer entries each at most n, then the size of A is $rc(log\ n)$ bits.*

Definition 5.3   Running Time: For a given input, the number of primitive operations or "steps" executed is called running time of the algorithm.

With increase in the number of operations, the running time of the algorithm increases. The running time of an algorithm is defined in best case, worst-case, and average-case.

1. **Best-Case Running Time:** The best-case running time of an algorithm is a lower bound on the running time for an input, expressed as a function of the input size *n*. It is the function defined by the minimum number of steps taken on any instance of size *n*.

2. **Worst-Case Running Time:** The worst-case Running time of an algorithm is an upper bound on the running time for an input, expressed as a function of the input size *n*. It is the function defined by the maximum number of steps taken on any instance of size *n*.

3. **Average-Case Running Time:** The average-case running time of an algorithm is the average running time over all inputs of a fixed size, expressed as a function of the input size *n*. It is the function defined by the average number of steps taken on any instance of size *n*.

Average-case time is often difficult to determine. We usually focus on the worst-case running time because it is easier to analyze for any algorithm.

5.2 Big-$\mathcal{O}$ Notation

The running time of an algorithm is a function of input size n. So to compare two algorithms with same input, we need to compare the growth rate of those functions. To compare the growth rate of the functions, we use the big-$\mathcal{O}$ (oh) notation. It is used to describe the asymptotic behavior of the functions. Basically, it tells that how fast a function grows or declines. The rate of growth of a function is also called its order. The big-$\mathcal{O}$ notation is also known as Landau's symbol after the name of German number theoretician Edmund Landau who invented the notation.

Definition 5.4 **Big-$\mathcal{O}$** Given real valued functions $f(n)$ and $g(n)$, we say that $f(n)$ is $\mathcal{O}(g(n))$ i.e. $f(n) = \mathcal{O}(g(n))$ if there are positive constants c and n_0 such that $f(n) \leq c\, g(n)$ for $n \geq n_0$.

Following are some examples of functions expressed in big-$\mathcal{O}$ notation that often appear in the study of algorithms. Here, c is any arbitrary constant and n is the input of the algorithm.

1. Constant $= \mathcal{O}(1)$

2. Logarithmic $= \mathcal{O}(log\, n)$

3. Polylogarithmic $= \mathcal{O}((log\, n)^c)$

4. Linear $= \mathcal{O}(n)$

5. Quadratic $= \mathcal{O}(n^2)$

6. Cubic $= \mathcal{O}(n^3)$

7. Polynomial $= \mathcal{O}(n^c)$

8. Exponential $= \mathcal{O}(c^n)$

Example 76 *Consider the linear function* $f(n) = 2n + 10$. *We claim that* $f(n) = \mathcal{O}(n)$.
To show this, let for some c and n, $2n + 10 \leq cn \Rightarrow (c-2)n \geq 10 \Rightarrow n \geq \frac{10}{(c-2)}$. For $c = 3$ and $n_0 = 10$, we see that $2n + 10 \leq cn \forall n \geq n_0$. Thus $2n + 10 = \mathcal{O}(n)$.

Example 77 *The function $n^2 \neq \mathcal{O}(n)$. As if $n^2 \leq cn \Rightarrow n \leq c$. This inequality does not satisfy for any constant c. Thus $n^2 \neq \mathcal{O}(n)$.*

5.2.1 Big-$\mathcal{O}$ and Growth Rate

The big-$\mathcal{O}$ notation gives an upper bound on the growth rate of a function. The statement "$f(n)$ is $\mathcal{O}(g(n))$" means that the growth rate of $f(n)$ is not higher than the growth rate of $g(n)$. We can use the big-$\mathcal{O}$ notation to rank functions according to their growth rate. See the table below.

	$f(n)$ is $\mathcal{O}(g(n))$	$g(n)$ is $\mathcal{O}(f(n))$
$g(n)$ grows faster	Yes	No
$f(n)$ grows faster	No	Yes
Same growth	Yes	Yes

Following are some rules to write big-$\mathcal{O}$ notation.

■ If $f(n)$ is a polynomial of degree d, then $f(n) = \mathcal{O}(n^d)$ i.e., drop lower-order terms and constant factors.

■ Use the smallest possible class of functions e.g. "$2n = \mathcal{O}(n)$" instead of "$2n = \mathcal{O}(n^2)$".

■ Use the simplest expression of the class e.g. "$3n + 5 = \mathcal{O}(n)$" instead of "$3n + 5 = \mathcal{O}(3n)$".

In addition to big-$\mathcal{O}$ notation, we also define big-Omega Ω and big-Theta Θ notations below. These notations are also called order notation.

Definition 5.5　**Big-Omega:** Given real valued functions $f(n)$ and $g(n)$, we say that $f(n)$ is $\Omega(g(n))$ i.e. $f(n) = \Omega(g(n))$ if there is a positive constant c and an integer constant $n_0 \geq 1$ such that $f(n) \geq cg(n)$ for $n \geq n_0$.

Definition 5.6　**Big-Theta:** Given real valued functions $f(n)$ and $g(n)$, we say that $f(n)$ is $\Theta(g(n))$ i.e. $f(n) = \Theta(g(n))$ if there are constants $c' > 0$ and $c'' > 0$ and an integer constant $n_0 \geq 1$ such that $c'g(n) \leq f(n) \leq c''g(n)$ for $n \geq n_0$.

Thus, $f(n) = \mathcal{O}(g(n))$ if $f(n)$ is asymptotically less than or equal to $g(n)$, $f(n) = \Omega(g(n))$ if $f(n)$ is asymptotically greater than or equal to $g(n)$ and $f(n) = \Theta(g(n))$ if $f(n)$ is asymptotically equal to $g(n)$. See the following examples.

Example 78 *Let $f(n) = 7n^2$, then for $c = 7$ and $n_0 = 1$ we get $f(n)\Theta(n^2)$.*

5.2.2　Properties of Order Notation

For functions $f(n)$, $g(n)$, $h(n)$, and $l(n)$, the following are true:

1. $f(n) = \mathcal{O}(g(n))$ if and only if $g(n) = \Omega(f(n))$.

2. $f(n) = \Theta(g(n))$ if and only if $f(n) = \mathcal{O}(g(n))$ and $f(n) = \Omega(g(n))$.

3. If $f(n) = \mathcal{O}(h(n))$ and $g(n) = \mathcal{O}(h(n))$, then $(f + g)(n) = \mathcal{O}(h(n))$.

4. If $f(n) = \mathcal{O}(h(n))$ and $g(n) = \mathcal{O}(l(n))$, then $(f.g)(n) = \mathcal{O}(h(n)l(n))$.

5. (reflexivity) $f(n) = \mathcal{O}(f(n))$.

6. (transitivity) If $f(n) = \mathcal{O}(g(n))$ and $g(n) = \mathcal{O}(h(n))$, then $f(n) = \mathcal{O}(h(n))$.

Approximations of some commonly occurring functions:

■ Polynomial function: If $f(n)$ is a polynomial of degree k with positive leading term, then $f(n) = \mathcal{O}(n^k)$.

■ For any constant $c > 0$, $log_c\ n = \mathcal{O}(log\ n)$.

■ $log(n!) = \mathcal{O}(nlogn)$.

Comparative growth rates of some functions: [117] Let ε and c be arbitrary constants with $0 < \varepsilon < 1 < c$, then $1 < log(logn) < logn < exp\sqrt{p\ lg\ n(loglogn)} < n^\varepsilon < n^c < n^{logn} < c^n < n^n < c^{c^n}$.

Time Estimation of Mathematical Operations: Let m and n be two numbers of bit length k and l respectively. Then we write, $k = \mathcal{O}(log\ m), l = \mathcal{O}(log\ n)$.

■ **Time for Addition/Subtraction**

Time$(kbits + lbits) = max(k,l)$, so Time $(m+n) = O(max(log\ m, log\ n))$.

■ **Time for Multiplication**

For the multiplication of a k-bit number by an l-bit number, all the following statements are correct:

1. Time $= \mathcal{O}(kl)$

2. Time $< kl$

3. Time $= k(l-1)$

4. If the second number (having l bits) has equal number of 0-bits and 1-bits, then Time $= \frac{kl}{2}$. In general, we write in Big-$\mathcal{O}$ notation as Time $(m \times n) = \mathcal{O}(logn.logm) = \mathcal{O}(kl)$. In particular, if $m = n$ with k-bits length then Time $(kbits \times lbits) = \mathcal{O}(k^2)$.

Example 79 *Estimate the time required to compute n!*
Solution. *We want to compute $n! = 1 \times 2 \times 3 \times 4....n-1 \times n$ i.e. in $j - 1^{th}$ step we are multiplying $j!$ by $j+1$. We have $n-2$ multiplications, where each multiplication involves multiplying a partial product (namely $j!$) by the next integer. As a worst case estimate, let the number of binary digit in the last product in each multiplication be $n!$ (length $n! = \mathcal{O}(nlogn)$). So, in each of $n-2$ multiplication in the computation of $n!$, we are multiplying integer with at most $\mathcal{O}(logn)$ bits (namely, $j+1$) by an integer with $O(nlogn)$ (namely, $j!$). This requires $\mathcal{O}(nlog^2n)$ bit operations. We must do this $n-2 = O(n)$ times. Therefore, the total number of bit operations is $\mathcal{O}(nlog^2n) \times \mathcal{O}(n) = \mathcal{O}(n^2log^2n)$.*

Modular Exponentiation Modular exponentiation is a general method to find x^y mod n which is also known as 'square and multiply' method. The method is explained in detail in Chapter 3, Section 3.8.1.

Time Complexity of Modular Exponentiation: We can find the least non-negative residue of x^y (mod n) in time $\mathcal{O}(k^2 l)$, where n is a k-bit natural number, y is an l-bit natural number, and $|x| < n$.

For this, we first write y in binary expression as $y = e_{l-1} 2^{l-1} + e_{l-2} 2^{l-2} + ... + e_1 2 + e_0$. Clearly, $x^y = x^{e_{l-1} 2^{l-1} + e_{l-2} 2^{l-2} + ... + e_1 2 + e_0}$ mod n. We then successively compute x^{2^j} mod n for $j = 1...l-1$. To compute x^{2^j} mod n, we take the value $x^{2^{j-1}}$ mod n and just square it applying modulo n i.e. $x^{2^j} \equiv \{x^{2^{j-1}}\}^2$ mod n. Since size of each x^{2^j} mod n is at most k bits, so this process takes time $\mathcal{O}(k^2)$ for each j. The binary expression of y can have at most l 1-bits. Therefore, to get the result, we need at most l multiplications. Clearly, time for repeated squaring algorithm is $\mathcal{O}(k^2 l)$.

Example 80 *Compute* 2^{14} (mod 21).

Solution

$$
\begin{aligned}
2^{14} \quad (\text{mod } 21) = \quad &= \quad \text{mod } \_exp(2, 14, 21) \\
&= \quad \text{mod } \_exp(4, 7, 21) \\
&= \quad 4 \times \quad \text{mod } \_exp(16(= 4^2 \quad \text{mod } 21), 3, 21) \\
&= \quad 4 \times 16 \times \quad \text{mod } \_exp(4(= 16^2 \quad \text{mod } 21), 1, 21) \\
&= \quad 4 \times 16 \times 4 \times \quad \text{mod } \_exp(16(= 4^2 \quad \text{mod } 21), 0, 21) \\
&= \quad 4 \times 16 \times 4 \times \quad (\text{mod } 21) \\
&= \quad 4
\end{aligned}
$$

In fact $14 = 2^3 + 2^2 + 2$, *thus* $2^{14} \equiv 2^{2^3 + 2^2 + 2} \equiv 2^2 2^{2^2} 2^{2^3} \equiv 2^2 ((2^2)^2)(((2^2)^2))^2$ mod $21 = 4 \times 16 \times 4 \equiv 4$ (mod 21).

Some other important algorithms such as Euclidean Algorithm, Extended Euclidean Algorithm, primality testing algorithm, algorithms for solving factoring problem, and discrete logarithm problem with their time complexity are discussed in the Chapter 3 (Number Theory).

5.3   Types of Algorithm

On the basis of running time complexity, we categorize the algorithms in the following classes. For the below division, we assume the input size of the algorithm as k.

Definition 5.7   Polynomial Time Algorithm: If there exists a positive constant c, such that the running time of the algorithm is $\mathcal{O}(k^c)$, then the algorithm is called polynomial time algorithm.

Definition 5.8   Exponential Time Algorithm: If there exists a positive constant c, such that the running time of the algorithm is $\mathcal{O}(e^{ck})$, then the algorithm is called exponential time algorithm.

Definition 5.9   Sub-exponential Time Algorithm: If the running time of the algorithm is between polynomial time and exponential time i.e. the algorithm is slower than any polynomial but is still significantly faster than exponential, then the algorithm is called sub-exponential time algorithm.

Let n, γ, c be real numbers where $n > e$ (*i.e.* 2.718). Write $L_n[\gamma, c] = \mathcal{O}(e^{c(\log n)^{\gamma}(\log \log n)^{1-\gamma}})$. Then, we have $L_n[0, c] = \mathcal{O}((\log n)^c)$, a polynomial time and $L_n[1, c] = \mathcal{O}(e^{c(\log n)})$, an exponential time. The algorithm having running time $L_n[\gamma, c], 0 < \gamma < 1$ is called a sub-exponential time algorithm.

Definition 5.10   Deterministic Algorithm: If for a given particular input, the algorithm always produces the same unique output, then the algorithm is called a deterministic algorithm. The output of the algorithm depends only on the input. Deterministic algorithms follow the same execution path (sequence of operations) each time they are executed with the same input.

Definition 5.11   Randomized Algorithm: If the output of the algorithm depends not only on the given input but also on some random number chosen at certain points in the execution path of the algorithm, then the algorithm is called a randomized algorithm. So, for the same input, the algorithm may produce different outputs.

Deterministic algorithms are not practically efficient. However, randomized algorithms are practically efficient.

5.4   Complexity Classes

In this section, we will present only the formal definitions of some complexity classes. For details about the complexity theory and complexity classes, the reader may refer to some standard books on complexity theory such as [64, 5].

An algorithm is designed to solve a problem. Any task given to solve is called a problem, and a particular case of the task is called an instance of the problem. For example, if we are given to factor any composite number, then factoring is a problem, but if we want to factor a particular composite number then it is an instance of factoring problem. We can classify all the problems in two classes, computational/search problem and decision problem.

Definition 5.12   Decision Problem: If the solution (output) of a problem depends only on the decision between YES or NO, then the problem is called decisional problem. For example, to decide whether the given number is composite or not is a decisional problem.

Definition 5.13   Computational Problem: If the desired solution does not depend only on deciding YES or NO but require some computations too, then the problem is called computational/search problem. For example, the problem to test whether the number is composite or not, if yes then find its factors also.

Remark 5.1   All the computational problems can be phrased as decisional problems in such a way that an efficient algorithm for the decisional problem yields an efficient algorithm for the computational problem and vice versa.

Definition 5.14   Complexity Class *P*: It is the set of all decisional problems that are solvable in polynomial time.

In the solution of any decisional problem, the output of the algorithm is either YES or NO. We can verify the YES/NO in polynomial time algorithm. On the basis of this verification, again we can classify the problem in the below two classes.

Definition 5.15   Complexity Class NP: It is the set of all decisional problems for which YES answer can be verified in polynomial time given some extra information, called certificate.

Definition 5.16   Complexity Class Co-NP: It is the set of all decisional problems for which NO answer can be verified in polynomial time given some extra information, called certificate.

Example 81 *Let the decisional problem be "COMPOSITES" i.e. to decide whether the given number "N" is composite or not. The "COMPOSITES" belongs to the complexity class NP, because if the input number "N" is composite, then dividing "N" by one of its divisor a (the certificate), we can verify the YES answer in polynomial time. It also belongs to the Co-NP class. But it is still an open problem to test whether "COMPOSITES belongs to class P".*

Remark 5.2   If a decisional problem is in NP, then getting certificate of YES answer may not be easy to find, but if the certificate is obtained and is known, then it can be used to verify the YES answer in polynomial time. The same is true for the case in Co-NP.

Remark 5.3 $P \subset NP$ and $P \subset Co - NP$.

Remark 5.4 Some open problems in complexity theory are (i) Is $P = NP$? (ii) Is $NP = Co - NP$? (iii) Is $P = NP \cap Co - NP$?.

Definition 5.17 Polynomial Time Reduction: Let A_1 and A_2 be two decisional problems. We say that A_1 reduces into A_2 in polynomial time, if there exists a polynomial time algorithm (as a function of the A_1) which can construct an instance I_2 of A_2, given any instance I_1 of A_1, such that the answer for I_1 is same as the answer for I_2. In this case, we write $A_1 \leq_P A_2$.

Remark 5.5 If $A_1 \leq_P A_2$, then A_2 is at least as difficult as A_1 or, equivalently, A_1 is no harder than A_2.

What is the importance of polynomial time reduction? Suppose, we have an efficient algorithm for A_2, if A_1 gets reduced to A_2 in polynomial time, then we can use the algorithm for A_2 to solve A_1 as well.

Remark 5.6 In the above case, if there does not exist an efficient algorithm for A_1, then there is no efficient algorithm for A_2 also.

Definition 5.18 Computationally Equivalent: Let A_1 and A_2 be two decisional problems. If $A_1 \leq_P A_2$ and $A_2 \leq_P A_1$, then A_1 and A_2 are said to be computationally equivalent.

Definition 5.19 NP-**Complete (NPC) Problem:** A decisional problem A is said to be NP-complete if : (i) $A \in NP$ and (ii) $A_1 \leq_P A$ for every $A_1 \in NP$.

Remark 5.7 NP-complete problems are the hardest problems in the NP class, in the sense that they are at least as difficult as every other problem in the NP class.

Definition 5.20 NP-**Hard Problem:** A problem (decision or computational/search) is NP-hard if there exists some NP-complete problem that can be polynomial time reduces to it i.e. if there is an algorithm to solve the problem, then that algorithm can be translated into another algorithm to solve some NP complete problem (polynomial time problem).

Remark 5.8 The NP-hard classification is not restricted only to decisional problems.

Remark 5.9 NP-complete problem is also NP-hard.

Definition 5.21   The Complexity Class PP: A problem A is said to belongs to class *PP* if there exists a probabilistic polynomial time algorithm $\mathcal{A}$ such that $\mathcal{A}$ recognizes any instance $I \in A$ with certain error probability.

The error probability is formulated as below:

$$Prob[\mathcal{A} \ recognizes \ I | I \in A] \geq \varepsilon \qquad (5.1)$$

$$Prob[\mathcal{A} \ recognizes \ I | I \notin A] \leq \delta \qquad (5.2)$$

Where ε and δ are constants satisfying $\varepsilon \in (1/2, 1]$ and $\delta \in [0, 1/2)$.

The expression in the first probability bound (Equation (5.1)) is for a correct recognition of an instance. It is called the completeness probability bound. The expression in the second probability bound (Equation (5.2)) is for a mistaken recognition of a non-instance. It is called soundness probability bound. If $\delta = 0$ and $\varepsilon = 1$, then there is no error at all. Algorithm in such class of *PP* is called zero side error *PP* algorithm (*ZPP*). If $\varepsilon = 1$ and $\delta \neq 0$, then it is called *PP* (Monte Carlo) algorithm. And if $\delta = 0$ and $\varepsilon \neq 1$, then the algorithm is called *PP* (Las-Vegas) algorithm.

Thus, we can define the Las-Vegas and Monte-Carlo algorithm as follows.

Definition 5.22   Las-Vegas algorithm: A randomized algorithm which may not give an answer but any answer it gives is always correct i.e. it always gives a correct answer or no answer at all. Sometimes this algorithm is called probably fast and always correct.

Definition 5.23   Monte Carlo Algorithm: A YES-based Monte Carlo algorithm is a randomized algorithm for a decisional problem in which "YES" answer is always correct, but "NO" answer may be incorrect. A NO-based Monte Carlo is defined in the similar way where "NO" answer is always correct but "YES" answer may be incorrect. Sometimes this algorithm is called always fast and probably correct.

We say that a YES-based Monte Carlo algorithm has an error probability equal to ε if, for any instance in which the answer is "YES", the algorithm will give the incorrect answer "NO" with the probability at most ε.

In Chapter 3 we have discussed some *PP* algorithms to test whether the given number is prime or not.

5.5   Exercises

1. Show that: (i) $5n - 2 = \mathcal{O}(n)$, (ii) $2n^3 + 10n^2 + 5 = \mathcal{O}(n^3)$, (iii) $3 \log n + 7 = \mathcal{O}(\log n)$.

2. Show that $7n^2 = \Theta(n^2)$.

3. Using big-$\mathcal{O}$ notation, estimate the time to compute n^n.

4. Using big-$\mathcal{O}$ notation, estimate the time to compute 2^n.

5. Consider two problems P_1 and P_2. Where P_1 is the problem, "If a quadratic polynomial $p(x)$ with integer coefficient is given, does $p(x)$ have two distinct real roots?". And the problem P_2 is "Given an integer N, is N positive?" Show that problem P_1 is computationally equivalent to problem P_2.

6. Consider two problems P_1 and P_2. Where P_1 is the problem "Given a polynomial $p(x)$ with integer coefficients, is there any interval of the real number line on which $p(x)$ decreases?. And the problem P_2 is "Given a polynomial $p(x)$ with integer coefficients, is there any interval of real number line on which $p(x)$ is negative". Prove that problem P_1 is polynomial time reducible to problem P_2.

7. Prove that the RSA problem is polynomial time reducible to factoring problem. RSA problem is to find x for given $x^e \mod n$, where $n = pq$ product of two distinct primes and $gcd(e, \phi(n)) = 1$.

8. Prove that the Diffie-Hellmann (DH) problem is polynomial time reducible to discrete logarithm problem (DLP), where DH problem is to find $g^{ab} \mod p$, for given $g^a \mod p$ and $g^b \mod p$, where p is a prime number and g is a generator of cyclic group Z_p^*. And the DLP is to find a for a given $g^a \mod p$ and p, where p is a prime number and g is a primitive root $\mod p$.

Chapter 6

Classical Cryptosystems

CONTENTS

In this chapter, we will discuss some simple cryptosystems popularly used in the pre-computer era. These are called classical cryptosystems. We will also discuss some possible attacks on those cryptosystems. In contrast to modern cryptographic algorithms, these cryptosystems can be solved easily. But they have their own importance as they have fulfilled the objective of secure communication in pre-machine era. Before describing the cryptosystems, we make some conventions.

- ■ Throughout this chapter we use lowercase letters for plaintext and uppercase letters for CIPHERTEXT.

- ■ To express a cryptosystem mathematically, the message is expressed in the form of numbers. For this, each letter of the alphabet is assigned a number as follows:

a	*b*	*c*	*d*	*e*	*f*	*g*	*h*	*i*	*j*	*k*	*l*	*m*
00	01	02	03	04	05	06	07	08	09	10	11	12
n	*o*	*p*	*q*	*r*	*s*	*t*	*u*	*v*	*w*	*x*	*y*	*z*
13	14	15	16	17	18	19	20	21	22	23	24	25

all the mathematical operations are performed under modulo 26.

- ■ Usually, we have written the messages without space and punctuation. To use space, punctuation, or any special symbol we represent them by some numbers (like 26, 27, 28...) accordingly. And in this case, the mathematical operation under modulo 26 is replaced by modulo 27, modulo 28,...respectively.

- ■ The ideal scenario of attack is that there is a sender, a receiver, and meanwhile some attacker who wants to trap the confidential information shared between the users. We use the names Alice- for sender, Bob- for receiver, and Eve- for attacker.

6.1   Classification of Classical Cryptosystem

The word cipher is synonymous to cryptosystem. On the basis of the encryption process, we can classify the classical cryptosystems as follows:

1. Substitution Cipher

    (a) Mono-alphabetic substitution

        i. Shift Cipher
        ii. AFFINE Cipher
        iii. Substitution Cipher

(b) Poly-alphabetic substitution

    i. Vigenere Cipher

    ii. Hill Cipher

    iii. Auto-Key Cipher

2. Transposition Cipher

(a) Scytale

(b) The Rail Fence Cipher

Apart from the above classification, we can also classify all the cryptosystems into Block Cipher and Stream Cipher.

6.2 Block Cipher

In block cipher, the key and the encryption algorithm is applied to the block of data at once, instead of applying on each single alphabet/bit individually. There are two types of block cipher: Substitution Cipher and Transposition Cipher.

6.2.1 Substitution Cipher

As per its name, in this cipher the alphabets/characters are substituted by another alphabets/characters. The substituted alphabet may or may not belong to the given plaintext block. During the encryption, one character can be either replaced by a unique character or by more than one character. In the first case, the cipher is called to *mono-alphabetic cipher* and in the second case it is called *poly-alphabetic cipher*. In this section, we will discuss six classical cryptosystems, the first three belong to mono-alphabetic cipher and last three belong to poly-alphabetic cipher.

6.2.1.1 Shift Cipher

It is one of the earliest cryptosystems. In this cipher, the plaintext space $(\mathcal{P})$ = ciphertext space $(\mathcal{C})$ = key space $(\mathcal{K})$ = $\mathbb{Z}_{26}$. For any $k \in \mathcal{K}$ and $x \in \mathcal{P}$, the encryption and decryption process is defined as follows:

- Encryption : $E_k(x) = (x+k) \mod 26 = Y$.
  An encryption algorithm E_k substitutes any letter by the one occurring k positions ahead (cyclic) in the alphabet. y is the ciphertext corresponding to the plaintext x. Alice first repeats the process for each plaintext character, then sends the complete ciphertext to Bob.

- Decryption : $D_k(Y) = (Y-k) \mod 26 = x$.
  A decryption algorithm D_k substitutes any letter by the one occurring k positions backward (cyclic) in the alphabet. After receiving the ciphertext, Bob decrypts each character (Y) of ciphertext using decryption process.

Note that since the order of key space #($\mathcal{K}$) = 26, a very small number, hence, exhaustive key search attack (brute force attack) is possible in this cipher. In this attack, attacker try to use all possible keys one-by-one to decrypt the target ciphertext.

Example 82 Caesar Cipher *Caesar cipher, named after the Roman dictator Julius Caesar (100-44 B.C.), is a famous shift cipher where k = 3, i.e. in this cipher an alphabet is simply shifted by 3 places forward during encryption and 3 places backward during decryption.*
Mathematically, we can define encryption of Caesar cipher as
$E_k(x) = (x+3) \mod 26$
and the decryption as
$D_k(Y) = (Y-3) \mod 26.$

Consider the following example,

Plaintext = i came i saw i conquered
Ciphertext = L FDPH L VDZ L FRQTXHUHG

Example 83 *Some particular shift ciphers are as below:*
$E_2(example) = GZCORNG,$
$E_3(example) = HADPSOH,$
$E_1(hal) = IBM,$
$E_3(cold) = FROG.$

We observe that the shift cipher (modulo 26) is not secure due to the small key size. It can be analyzed by exhaustive key search attack (brute-force attack). Since there are only 26 possible different keys, it is easy to decrypt the ciphertext using all possible keys until we get a "meaningful" plaintext string. This is explained in the following example.

Example 84 *Let the given ciphertext string be "MJAIAMWLXSVITPEGIPIXXIVW".*

Apply the following trials using keys 0,1,2,3...etc.

Trial 1 : lizhzlvkwruhsodfhohwwwhuv (Shift backward 1)
Trial 2 : khygykujvotgrncegngvvgtu (Shift backward 2)
Trial 3 : jgxfxjtiupsfombdfmfuufst (Shift backward 3)
Trial 4 : ifwewishtoreplaceletters (Shift backward 4)

So the plaintext is: "if we wish to replace letters".

The major problem with shift ciphers is "the key space is too small against brute-force attack". As the above example indicates, a necessary condition for a secure cryptosystem is that an exhaustive key search should be infeasible i.e. the key space should be very large. However, very large key space is not sufficient for the guarantee of a secure cryptosystem.

6.2.1.2 Affine Cipher

Here also, the plaintext space $(\mathcal{P})$ = ciphertext space $(\mathcal{C})$ = $\mathbb{Z}_{26}$, but the key space is $\mathcal{K} = \{(a,b) \in \mathbb{Z}_{26} \times \mathbb{Z}_{26} :$ where $gcd(a,26) = 1\}$. For any $k = (a,b) \in \mathcal{K}$ and $x \in \mathcal{P}$, the encryption and decryption process are as follows:

- Encryption: $E_{(a,b)}(x) = (ax+b) \mod 26 = Y$
  An encryption algorithm E_k substitutes any alphabet x by the image under the liner (affine) mapping $ax+b$, hence, the cryptosystem is called affine cipher. Y is the ciphertext corresponding to the plaintext x. Alice repeats the process for each plaintext character and sends the complete ciphertext to Bob.

- Decryption: $D_{(a,b)}(Y) = a^{-1}(Y-b) \mod 26 = x$
  A decryption algorithm D_k substitutes any alphabet by its inverse under the linear mapping $(ax+b) \mod 26$. After receiving the ciphertext, Bob decrypts each character (Y) of ciphertext using the decryption process.

The number of integers less than and relatively prime to m is given by $\phi(m)$ (Euler's phi function). So, the order of key space $\#(\mathcal{K})$ = possible values of $b(=26) \times$ possible values of $a(=12) = 312$, not a big number! Exhaustive key search attack is possible. Hence, this cipher is also not secure, but is more secure in comparison to shift cipher. Number of keys in the affine cipher over $\mathbb{Z}_m$ is $m\phi(m)$, if $m = 60$ then $\#\mathcal{K} = 60 \times 16 = 960$.

Example 85 *Consider the key $k = (9,2)$ and the message "affine".*

- *Encryption: The numerical value of the message "affine" is "000505081304".* *Since each alphabet is a plaintext, hence, six plaintexts in this message. Encrypting one by one, using the mapping $9x+2 \pmod{26}$, where $x = 0,5,5,8,13,4$, we obtain 022121221512. Thus, the ciphertext is "CVVWPM".*

- *Decryption: To decrypt "CVVWPM", first we convert the ciphertext into corresponding numerical value "022121221512". Now, decrypting the alphabets using the mapping $9^{-1}(Y-2) \equiv 3(Y-2) \equiv 3Y-6 \equiv 3Y+20 \pmod{26}$, where $Y = 02,21,21,22,15,12$, we obtain "000505081304". Thus, the original message is "affine".*

Why is the condition $gcd(a,26) = 1$ important? The answer is given in the following theorem.

Theorem 6.1
The congruence $ax \equiv b \pmod{m}$ has a unique solution $x \in \mathbb{Z}_m$ for every $b \in \mathbb{Z}_m$ if and only if $gcd(a,m) = 1$.

If a is not relatively prime to 26, then the equation $ax+b = y \pmod{26}$ may have more than one solutions. We can explain this by an example. Suppose the key is

$(13, 4)$, hence, the encryption rule is $13x + 4 \pmod{26}$. If we encrypt the plaintext "input", we get the ciphertext "ERRER" and if we encrypt the plaintext "alter", we get the same ciphertext "ERRER". Thus, encryption of two different plaintexts yields the same ciphertext. Encryption is not one-to-one, hence, not a valid encryption. If we decrypt "ERRER", we can get two plaintexts. Observe that $gcd(13, 26) \neq 1$.

6.2.1.3 Substitution Cipher

plaintext space $(\mathcal{P})$ = ciphertext space $(\mathcal{C})$ = $\mathbb{Z}_{26}$ and $\mathcal{K} = S_{26}$ (set of all permutations $\pi : \mathbb{Z}_{26} \to \mathbb{Z}_{26}$). For each permutation mapping π, we define

■ Encryption: $E_\pi(x) = \pi(x)$.

■ Decryption: $D_\pi(Y) = \pi^{-1}(Y)$.

where π^{-1} is the inverse permutation to π.

Example 86 *Let Alice and Bob agree on a permutation $\pi = (1, 2, 3, 4, ...25)$. And let plaintext be $x = ramacceptedbribe$.*

■ *Encryption: The numerical value of the message is* $= (1700120002020415190$ $4030107080104)$, *thus* $E_\pi(x) = \pi(1700120002020415190 4030107080104) = (1801130103030516200504 0208090205) = SBNBDDFQUFECSJCF.$

■ *Decryption: Using the inverse of permutation π, Bob decrypt the ciphertext "SBNBDDFQUFECSJCF" as* $D_\pi(Y) = \pi^{-1}(18011301030305162005040 20 8090205) = (1700120002020415190403010 7080104) = ramacceptedbribe.$

Key space of substitution cipher includes all possible permutations of 26 alphabets. Clearly, the order of the key space $\#\mathcal{K} = 26! > 10^{26}$ is a very large number, therefore, one cannot check case by case even using a computer. So, exhaustive search attack is not possible in this cipher. However, we will see later in cryptanalysis section that this cipher is not secure against some attacks, viz. frequency analysis. The shift cipher and affine cipher are special cases of substitution cipher.

In all of the above three cryptosystems; shift cipher, affine cipher, and substitution cipher, a single alphabet is replaced by a unique alphabet in encryption. These are called *mono-alphabetic substitution* and can be easily solved using *frequency analysis* (discussed in the next section). If a single alphabet of plaintext is replaced by more than one alphabet (*poly-alphabetic substitution*), then the cryptosystem may be more secure than mono-alphabetic cipher. Below, we discuss some substitution ciphers based on poly-alphabetic substitution.

6.2.1.4 Vigenere Cipher

The well-known Vigenere cipher is named after Blaise de Vigenere, a French diplomat and cryptographer who lived in the sixteenth century (1523–1596). It is a simple form of poly-alphabetic substitution.

In this cryptosystem, $\mathcal{P} = \mathcal{C} = \mathcal{K} = \mathbb{Z}_{26}^m$. For a key $K = (k_1, k_2, k_3, \ldots k_m) \in \mathcal{K}$ and a plaintext $x = (x_1, x_2, x_3 \ldots x_m) \in \mathcal{P}$, we define the encryption and decryption as follows:

- Encryption: $E_K(x_1, x_2, x_3, \ldots x_m) = (x_1 + k_1, x_2 + k_2, \ldots x_m + k_m) \mod 26$

- Decryption: $D_K(y_1, y_2, y_3, \ldots y_m) = (y_1 - k_1, y_2 - k_2, \ldots, y_m - k_m) \mod 26$

In Vigenere cipher, we encrypt/decrypt a collection of m-alphabetic plaintext/ciphertext using a string of m-alphabetic key called "keyword". It is explained by the following simple examples.

Example 87 *Let the keyword be "CODE", number of alphabet in keyword is 4, so $m = 4$. Suppose Alice wants to send a plaintext "the germans are coming" to Bob. Before communicating, they first share the keyword. Using the correspondence $A \leftrightarrow 0$, $B \leftrightarrow 1$, $C \leftrightarrow 2 \ldots Z \leftrightarrow 25$ as described earlier. She first obtains the numerical equivalent of keyword "CODE" and plaintext "the germans are coming". To encrypt the plaintext she does addition of the plaintext and key using residue modulo 26 as follows:*

Numeric value of the keyword is $(2, 14, 3, 4)$.

PT	t	h	e	g	e	r	m	a	n	s
numerical PT	19	7	4	6	4	17	12	0	13	18
Key	2	14	3	4	2	14	3	4	2	14
Encryption	21	21	7	10	6	5	15	4	15	6
CT	V	V	H	K	G	F	P	E	P	G
PT	a	r	e	c	o	m	i	n	g	—
numerical PT	0	17	4	2	14	12	8	13	6	—
Key	3	4	2	14	3	4	2	14	3	—
Encryption	3	21	6	16	17	16	10	1	9	—
CT	D	V	G	Q	R	Q	K	B	J	—

Thus, the ciphertext is "VVH KGFPEPG DVG QRQKBJ".
To decrypt the ciphertext, Bob does the above process in reverse order as below:

CT	V	V	H	K	G	F	P	E	P	G
Numerical CT	21	21	7	10	6	5	15	4	15	6
Key	2	14	3	4	2	14	3	4	2	14
Decryption	19	7	4	6	4	17	12	0	13	18
PT	t	h	e	g	e	r	m	a	n	s
CT	D	V	G	Q	R	Q	K	B	J	—
Numerical CT	3	21	6	16	17	16	10	1	9	—
Key	3	4	2	14	3	4	2	14	3	—
Decryption	0	17	4	2	14	12	8	13	6	—
PT	a	r	e	c	o	m	i	n	g	—

Thus, the plaintext is "the germans are coming".

Example 88 *Consider the keyword "CIPHER" and the message "thiscryptosys-temisnotsecure". The numerical value of keyword "CIPHER" is (2, 8, 15, 7, 4, 17). Using the process given in the previous example, we can encrypt the plaintext and get the ciphetext "VPXZGIZXIVWPUBTTMJPWIZITWZT".*

6.2.1.5 Hill Cipher

Hill cipher is a polygraphic substitution cipher based on linear algebra. The Hill cipher was invented by Lester S. Hill in 1929. This cryptosystem was probably never used. However, it has an important value in the study of historical cryptography. The Advanced Encryption Standard (AES) used the concept of Hill cipher to achieve the security criteria by using Diffusion and Confusion, proposed by Claude Shannon. The idea behind this cryptosystem is to take m linear combinations of m alphabetic characters in one (plaintext) block and produce m alphabetic characters in one (ciphertext) block.

In this cryptosystem, $\mathcal{P} = \mathcal{C} = \mathbb{Z}_{26}^m$ and the key space $\mathcal{K} = \{all\ m \times m\ invertible\ matrices\ in\ \mathbb{Z}_{26}\}$. To encrypt any message M, we first divide the message into finite number of blocks where each block is a plaintext containing m characters. We add some dummy characters Z or X if the last block is not of length m. For any key $K \in \mathcal{K}$ and plaintext $x = (x_1, x_2, ...x_m) \in \mathcal{P}$, the encryption $Y = E_K(x) = (y_1, y_2, y_3...y_m)$ and decryption $x = D_K(Y) = (x_1, x_2, ...x_m)$ is defined as follows:

- Encryption:

$$E_K(x_1, x_2, ...x_m) = (x_1, x_2, ...x_m) \begin{pmatrix} k_{1,1} & k_{1,2} & ... & ... & k_{1,m} \\ k_{2,1} & k_{2,2} & ... & ... & k_{2,m} \\ ... & ... & & & ... \\ ... & ... & & & ... \\ k_{m-1,1} & k_{m-1,2} & ... & ... & k_{m-1,m} \\ k_{m,1} & k_{m,2} & ... & ... & k_{m,m} \end{pmatrix}$$

- Decryption:

$$D_K(y_1, y_2, y_3...y_m) = (y_1, y_2, y_3...y_m) \begin{pmatrix} k_{1,1} & k_{1,2} & ... & ... & k_{1,m} \\ k_{2,1} & k_{2,2} & ... & ... & k_{2,m} \\ ... & ... & & & ... \\ ... & ... & & & ... \\ k_{m-1,1} & k_{m-1,2} & ... & ... & k_{m-1,m} \\ k_{m,1} & k_{m,2} & ... & ... & k_{m,m} \end{pmatrix}^{-1}$$

All the operations are performed under modulo 26. Let us consider the following simple example.

Example 89 *Suppose, the key is* $K = \begin{pmatrix} 11 & 8 \\ 3 & 7 \end{pmatrix}$.

Since $|K| \mod 26 = 1 \neq 0$. *The matrix K is invertible. By simple calculation, we have*

$$K^{-1} = \begin{pmatrix} 7 & 18 \\ 23 & 11 \end{pmatrix}.$$

Alice and Bob first share the secret key K. Suppose the message is "july". The numerical equivalent of "july" is $(9, 20, 11, 24)$. *Alice has two plaintext elements* $(9, 20)$ *(corresponding to "ju") and* $(11, 24)$ *(corresponding to "ly"). To encrypt the message "july", Alice computes under modulo 26 as follows:*

$$(9, 20) \begin{pmatrix} 11 & 8 \\ 3 & 7 \end{pmatrix} = (99 + 60, 72 + 140) = (3, 4)$$

and

$$(11, 24) \begin{pmatrix} 11 & 8 \\ 3 & 7 \end{pmatrix} = (121 + 72, 88 + 168) = (11, 22)$$

Thus, the encryption of $(9, 20, 11, 24)$ *is* $(3, 4, 11, 22)$. *Hence, the ciphertext is "DELW". To decrypt the ciphertext "DELW", Bob computes:*

$$(3, 4) \begin{pmatrix} 7 & 18 \\ 23 & 11 \end{pmatrix} = (21 + 92, 54 + 44) = (9, 20)$$

and

$$(11, 22) \begin{pmatrix} 7 & 18 \\ 23 & 11 \end{pmatrix} = (77 + 506, 198 + 242) = (11, 24).$$

Thus, the correct plaintext $(9, 20, 11, 24)$ = *"july" is obtained.*

In all the cryptosystems discussed above, a character of a plaintext is substituted by another character(s). Hence, they are called substitution cipher. Now, we will describe some cryptosystems in the next section where characters of a plaintext are not substituted by another character, but permuted inside the plaintext. Such ciphers are known as Transposition Cipher.

6.2.2   Transposition Cipher

In this cipher, the alphabets/characters of a given message (plaintext) are permuted to obtain ciphertext. This cipher is also called Permutation Cipher. The idea of permutation cipher is to keep the plaintext character unchanged but effect a change in their position using some permutation. A permution of a set X is a bijective mapping $\pi : X \rightarrow X$.

In this cryptosystem, $\mathcal{P} = \mathcal{C} = \mathbb{Z}_{26}^m$ and $\mathcal{K} = \{All \ permutation \ mapping \ \pi : \{1, 2, 3, ...m\} \rightarrow \{1, 2, 3, ...m\}\}$. Let $x = (x_1, x_2, ...x_m)$ be any plaintext element. The encryption and decryption is performed as follows:

■ Encryption: $E_{\pi}(x_1, x_2, x_3, ...x_m) = (x_{\pi(1)}, x_{\pi(2)}, x_{\pi(3)}, ..., x_{\pi(m)})$

■ Decryption: $D_{\pi}(y_1, y_2, y_3, ..., y_m) = (y_{\pi^{-1}(1)}, x_{\pi^{-1}(2)}, x_{\pi^{-1}(3)}, ..., x_{\pi^{-1}(m)})$

Consider the following example.

Example 90 *Let* $m = 3$ *and the key* $\pi = \begin{pmatrix} 1 & 2 & 3 \\ 2 & 3 & 1 \end{pmatrix}$. *The inverse of the permuta-*

tion mapping is $\pi^{-1} = \begin{pmatrix} 1 & 2 & 3 \\ 3 & 1 & 2 \end{pmatrix}$. *Let the plaintext be "london".*

To encrypt "london", Alice first divides message into two plaintext blocks "lon" and "don" of length 3. Now, she computes the ciphertext as below:

$E_\pi(lon) = ONL$ *and* $E_\pi(don) = OND$.

Thus, the ciphertext is ONLOND.

To decrypt the ciphertext "ONLOND", Bob proceeds as follows:

$D_\pi(ONL) = lon$ *and* $D_\pi(OND) = don$.

Thus, the plaintext is "london".

Below, we have described the Hill cipher as a poly-alphabetic substitution cipher. It can be easily shown that the Permutation Cipher is a special case of Hill Cipher.

Permutation Cipher as a Special Case of Hill Cipher: Consider a Permutation Cipher with the permutation π of the set $(1, 2, 3...m)$. For this permutation mapping, define a $m \times m$ permutation matrix $K_\pi = (k_{i,j})$ (in which every row and column contains exactly one "1" and all other entries are "0") as below:

$$k_{i,j} = \begin{cases} 1, & if \ i = \pi(j), \\ 0, & otherwise \end{cases}$$

It can be easily shown that $(K_\pi)^{-1} = K_{\pi^{-1}}$ and the Hill Cipher with the key K_π is equivalent to the Permutation Cipher with key π.

Consider the Example 90 discussed earlier. We can find the permutation matrix as below:

$$K_\pi = \begin{pmatrix} 0 & 0 & 1 \\ 1 & 0 & 0 \\ 0 & 1 & 0 \end{pmatrix}$$

and

$$K_\pi^{-1} = \begin{pmatrix} 0 & 1 & 0 \\ 0 & 0 & 1 \\ 1 & 0 & 0 \end{pmatrix}.$$ It can be easily verified that $K_\pi^{-1}K_\pi = I$.

If we encrypt the plaintext "london" using this permutation matrix K_π, we get the same ciphertext "ONLOND" as obtained in Example 90.

6.2.2.1 Scytale

Scytale is one of the oldest cryptosystem. This technique was popularly used by Spartans for confidential political and defence communications. In this method, the sender and receiver both posses a cylinder called 'scytale' of exactly the same radius and length. The sender use to wound a narrow ribbon of parchment around a cylinder, then writes on it lengthwise and sends the unwounded ribbon to the indented receiver. The message in the unwounded ribbon can be read only by a person who has a cylinder of exactly the same circumference.

If the circumference of a cylinder is c units, then the idea of Scytale is to write plaintext into c rows with fixed symbols in each row and then read it by columns to get cryptotext. Consider the following example.

Example 91 *Let the circumference of the cylinder be $c = 5$. Let the plaintext be "cryptographyisveryuseful". There are 24 characters in the plaintext. It is sufficient to write 5 characters in a row to encrypt the plaintext as below:*

$$
\begin{array}{ccccc}
c & r & y & p & t \\
o & g & r & a & p \\
h & y & i & s & v \\
e & r & y & u & s \\
e & f & u & l &
\end{array}
$$

Thus, the ciphertext is "COHEERGYRFYPASUITPVS".

6.2.2.2 The Rail Fence Cipher

The rail fence cipher (also called a zigzag cipher) is a form of transposition cipher. It derives its name from the way in which it is encoded. The text is written in a zigzag manner in two or many rows and then is read row by row. Here, the secret is the number of rows. See the following example:

Example 92 *Let the plaintext be "attackrome" and the secret key is $k = 2$.*

Encryption: Write the message in a zigzag manner in two rows and read row by row as below:

$$
\begin{array}{ccccc}
a & t & c & r & m \\
t & a & k & o & e
\end{array}
$$

Thus, the ciphertext is "ATCRMTAKOE".

Decryption: To decrypt the ciphertext, Bob writes the ciphertext in two rows and reads column by column as below:

$$
\begin{array}{ccccc}
a & t & c & r & m \\
t & a & k & o & e
\end{array}
$$

Thus, the plaintext is "attackrome".

In the cryptosystems we have studied so far, to encrypt a message we first divide the message into a fixed length (say m) of plaintext block. The successive plaintext blocks are encrypted using the same key K. That is, for the given plaintext string $x = X_1 X_2 X_3 ... X_n$ where each plaintext X_i is a fixed length block of alphabets $X_i = x_{i,1} x_{i,2} x_{i,3} ... x_{i,m}$ (say). The corresponding ciphertext string Y is obtained as follows:

$Y = Y_1 Y_2 Y_3 .... Y_n = E_K(X_1) E_K(X_2) ... E_K(X_n)$.

Cryptosystem with this principle are called *Block Cipher*. More details of modern block ciphers are discussed in the Chapter 7. In the next section, we will study an alternative approach of cipher called Stream Cipher.

6.3 Stream Cipher

The idea of stream cipher is to use a key stream $z = z_1 z_2 z_3 ....$ to encrypt the given whole plaintext stream $x = x_1 x_2 x_3 .....$, applying the key bitwise as follows:

$$Y = y_1 y_2 y_3 .... = E_{z_1}(x_1) E_{z_2}(x_2) E_{z_3}(x_3) ...$$

We can generate the keystream using a keystream generator/algorithm. On the basis of keystream generation process, the stream cipher can be divided in two parts : Synchronous Stream Cipher and Non-Synchronous Stream Cipher.

6.3.1 Synchronous Stream Cipher

In this stream cipher, the keystream is generated from the key, independent of the plaintext string, using some special keystream generation algorithms. This cipher can be defined as follows:

A *synchronous stream cipher* is a tuple $(\mathcal{P}, \mathcal{C}, \mathcal{K}, \mathcal{L}, \mathcal{E}, \mathcal{D})$ together with a key-generation function g such that following conditions are satisfied:

1. $\mathcal{P}$ is a finite set of possible plaintexts.

2. $\mathcal{C}$ is a finite set of possible ciphertexts.

3. $\mathcal{K}$, the key space, is a finite set of possible keys.

4. $\mathcal{L}$ is a finite set called keystream alphabet.

5. g is a keystream generator, which, on input a key K, generates an infinite keystream $z_1 z_2 z_3 ...$, where $z_i \in \mathcal{L}, \forall i \geq 1$.

6. For each $z \in \mathcal{L}$, there is an encryption rule $E_z \in \mathcal{E}$ and a corresponding decryption rule $D_z \in \mathcal{D}$ such that $E_z : \mathcal{P} \to \mathcal{C}$ and $D_z : \mathcal{C} \to \mathcal{P}$ and $D_z(E_z(x)) = x$ for every plaintext element $x \in \mathcal{P}$.

To illustrate this definition, we describe the Vigenere cipher as a synchronous stream cipher.

6.3.1.1   Vigenere Cipher as Synchronous Stream Cipher

Let m be the keyword length of a Vigenere Cipher with secret key $(k_1, k_2, k_3..., k_m)$. We define $\mathcal{K} = \mathbb{Z}_{26}^m$ and $\mathcal{P} = \mathcal{C} = \mathcal{L} = \mathbb{Z}_{26}$. The keystream generator can be defined as follows:

$$z_i = \begin{cases} k_i & \text{if } 1 \leq i \leq m \\ z_{i-m} & \text{if } i \geq m+1 \end{cases}$$

Thus the keystream generator generates the keystream $k_1 k_2 k_3 ... k_m k_1 k_2 k_3 ... k_m k_1 k_2 .....$ from the given initial key $(k_1, k_2, k_3..., k_m)$.

Now, for the given plaintext string $x = x_1 x_2 .. x_m x_{m+1} ......$, the encryption and decryption works as follows:

■   Encryption: $E_z(x) = (x+z) \mod 26$

■   Decryption: $D_z(y) = (y-z) \mod 26$

where binary operations are performed component-wise.

Remark 6.1   Shift and Affine Cipher are the special cases of stream cipher where the keystream is constant : $z_i = k \; \forall i$.

Remark 6.2   In general, a block cipher is a special case of stream cipher where the keystream is the representation of the keyword of the block cipher.

Definition 6.1   *Periodic Stream Cipher*: A stream cipher is called periodic stream cipher with period d if $z_{i+d} = z_i \; \forall i \geq 1$.

Any block cipher can be defined as a periodic stream cipher with period d where d is size of secret key.

The modern stream ciphers are described in terms of binary alphabets i.e. $\mathcal{P} = \mathcal{C} = \mathcal{L} = \mathbb{Z}_2$ and the encryption and decryption operations are just addition modulo 2 :
$E_z(x) = (x+z) \mod 2$
and
$D_z(y) = (y+z) \mod 2$.

Note that, under modulo 2 the addition and subtraction both have the same meaning i.e. $(a-b) \mod 2 = (a+b) \mod 2$ for any $a, b \in \{0, 1\}$.

Nature of the Keystream: The security of the stream cipher completely depends on the key stream. A central requirement for the key stream bits is that it should look like a random sequence to an attacker. Otherwise, an attacker Eve could guess the bits of keystream and can decrypt the ciphertext. A random keystream can be generated using a random number generator.

6.3.2 Linear Feedback Shift Register (LFSR)

In the previous section, we have discussed how to generate the synchronous key stream using a block cipher i.e. Vigenere Cipher. There we have generated a keystream of period m, if m is the initial keyword length. Let us look another method to generate synchronous keystream using a linear recurrence relation. Here, we will work over binary alphabets. Using an initial keystring $(k_1, k_2, ...k_m)$ and a linear recurrence relation, we can generate a keystream $(z_1, z_2, z_3....)$ of higher period (i.e. period $> m$).

Let us consider the initial keystring (keyword) $(k_1, k_2, ...k_m)$ and the linear relation of degree m as:

$$z_{i+m} = \sum_{j=0}^{m-1} c_j z_{i+j} \pmod{2}, \forall i \geq 1, \text{ where } z_i = k_i, i = 1, 2, 3...m \text{ and } c_0, c_1, c_2, ... \in \mathbb{Z}_2 \text{ are predefined constants.}$$

Each term z_{i+m} of the keystream depends on previous m terms, hence, the degree of this recurrence is m. Also note, that $c_0 \neq 0$ otherwise the recurrence will be of degree at most $m - 1$. Thus, in the stream cipher, the secret key K consists of $2m$ values $k_1, k_2, k_3...k_m, c_0, c_1, c_2, ..., c_m$. With the help of these $2m$ values, users (Alice and Bob) can generate a keystream of a finite period greater than m to encrypt/decrypt the plaintext/ciphertext. Due to the finite number of recurring states, the output sequence of an LFSR repeats periodically. This is also illustrated in Example 93. Moreover, an LFSR can produce output sequences of different lengths, depending on the feedback coefficients. Since an m-bit state vector can only assume $2^m - 1$ nonzero states, the maximum sequence length before repetition is $2^m - 1$.

Theorem 6.2
The maximum sequence length (period) generated by an LFSR of degree m is $2^m - 1$.

Example 93 *Suppose $m = 3$, initial keystring is $(0, 0, 1)$ and the linear recurrence is $z_{j+3} = (z_j + z_{j+1}) \pmod{2}$. Thus, here $c_0 = 1, c_1 = 1, c_2 = 0$ and hence, the key $K = (0, 0, 1, 1, 1, 0)$. Using these initial values, we can generate the key stream 0010111001011110010111... as follows:*

$$z_4 = z_1 + z_2 \pmod{2} = 0$$
$$z_5 = z_2 + z_3 \pmod{2} = 1$$

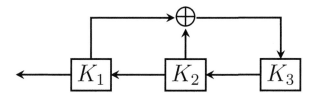

Figure 6.1: A Linear Feedback Shift Register

$z_6 = z_3 + z_4 \pmod 2 = 1$
$z_7 = z_4 + z_5 \pmod 2 = 0....etc$

Thus, the period of the keystream is $7(= 2^3 - 1)$. See Figure 6.1 to visualize the keystream generation process.

Example 94 *LFSR with maximum output sequence–Given an LFSR of degree $m = 4$, and the feedback path $(c_3 = 0, c_2 = 0, c_1 = 1, c_0 = 1)$, the output sequence of the LFSR has a period of $2^m - 1 = 15$ i.e. it is a maximum-length LFSR.*

Period of the keystream of LFSR depends on feedback path (recurrence relation). Consider the following example.

Example 95 *LFSR with non-maximum output sequence–Given an LFSR of degree $m = 4$ and $(c_3 = 1, c_2 = 1, c_1 = 1, c_0 = 1)$, then the output sequence has period of 5; therefore, it is not a maximum-length LFSR.*

Under what condition can we find a maximum period keystream in LFSR?
We can associate a polynomial to the given LFSR feedback (recurrence relation). For example, an LFSR with a feedback coefficient vector $(c_{m-1}, ..., c_1, c_0)$ is represented by the polynomial

$$P(x) = x^m + c_{m-1}x^{m-1} + ... + c_1 x + c_0$$

If the polynomial $p(x)$ is a *primitive polynomial*, then we get a maximum period LFSR keystream. For the definition of primitive polynomial the reader can refer to Chapter 2 of this book.

6.3.3 *Non-Synchronous Stream Cipher*

A *non-synchronous stream cipher* is a stream cipher in which each key stream element z_i depends on previous plaintext and/or ciphertext elements as well as the key K. A simple example of non-synchronous stream cipher is Autokey cipher.

6.3.3.1 Autokey Cipher

Autokey cipher is a type of stream cipher. An autokey cipher (also known as the autoclave cipher) is a cipher which incorporates the message (the plaintext) into the key, hence, the term autokey. In this cryptosystem, plaintext space $\mathcal{P}$ and ciphertext space $\mathcal{C}$ are streams of alphabet. The key space $\mathcal{K} = \mathbb{Z}_{26}$. Let the plaintext string is $x = x_1 x_2 x_3....$ Before communication, Alice and Bob first share the secret key $k \in \mathbb{Z}_{26}$. The encryption and decryption is defined as follows:

■ Encryption: To encrypt the plaintext, Alice generates a keystream $z_1, z_2, ...$ where $z_1 = k$ and $z_i = x_{i-1}, i \geq 2$.
She finds the ciphertext by computing $E_k(x) = (x_1 + z_1, x_2 + z_2, ...x_n + z_n...)$ mod 26

■ Decryption: After receiving the ciphertext $y = (y_1, y_2, y_3....)$, using the secret key k, Bob finds the corresponding plaintext by computing $D_k(y) = (y_1 - z_1, y_2 - z_2...y_n - z_n)$ mod 26.

Example 96 *Let the key $k = 8$, and the message be "attack rome".*

Plaintext (PT)	a	t	t	a	c	k	r	o	m	e
Numerical PT	0	19	19	0	2	10	17	14	12	4
Keystring	8	0	19	19	0	2	10	17	14	12
Encryption	8	19	12	19	2	12	1	5	0	16
Ciphertext	I	T	M	T	C	M	B	F	A	Q

Thus, the ciphertext is "ITMTCMBFAQ". After receiving the ciphertext, Bob can compute the plaintext as below:

Ciphertext	I	T	M	T	C	M	B	F	A	Q
Numerical CT	8	19	12	19	2	12	1	5	0	16
Keystrig	8	0	19	19	0	2	10	17	14	12
Decryption	0	19	19	0	2	10	17	14	12	4
Plaintext	a	t	t	a	c	k	r	o	m	e

Thus, the plaintext is "attack rome".

The autokey cipher can be extended by taking the key space $\mathcal{K} = \mathbb{Z}_{26}^m$ for some integer m. In this case, the shared secrete key is $K = (k_1, k_2, ...k_m)$, and then keystring can be generated as $z_i = k_i, 1 \leq i \leq m$, & $z_i = x_{i-m}, i > m$.

As there are attempts of attack on the designed cipher always, the designer of the cryptosystem must check the strength and weakness of the cryptosystem. This observation and computation, in order to, either attack on the cryptosystem or to check its strength against possible analysis is called cryptanalysis.

6.4   Cryptanalysis of Cryptosystems

In this section, we discuss some techniques of cryptanalysis of classical cryptosystems. The general assumption is that the cryptosystem follows the Kerckhoffs's principle i.e. the attacker Eve knows which cryptosystem is being used. If the attacker is unable to break the cryptosystem even when he knows about the cryptosystem, then the cryptosystem can be used in public network only. The types of attacks on cryptosystem are discussed in Chapter 1. We first consider here the weakest attack, the ciphertext only attack. The ciphertext only attack is possible on shift cipher, affine cipher, substitution cipher and Vigenere cipher. The Hill cipher and the LFSR stream cipher are secure against ciphertext only attack but vulnerable to known plaintext attack. The ciphertext only attack on classical block cipher are based on frequency analysis.

6.4.1   Frequency Analysis

Human languages are not random. Alphabets/letters of any human language are not uniformly distributed. For example, in English language, E is the most common letter, followed by T, A, O, I, N, S. Other letters such as Z, Q, X, J, K are fairly rare used to write any text. There are tables of single, double and triple letter (called diagram, trigram etc.) frequencies for various languages. Using statistical properties of English language, statistical distribution of 26 alphabets in numerous novels, magazines, and news papers, Beker and Piper obtained a frequency as given in Table 6.1 and Figure 6.2. On the basis of the statistical/probabilistic distribution of 26 letters alphabets, Beker and Piper divided these letters into five groups as follows:

1. E, having probability about 0.120.

2. T,A,O,I,N,S,H,R, each having probability between 0.06 and 0.09.

3. D, L, each having probability around 0.04.

4. C,U,M,W,F,G,Y,P,B, each having probability between 0.015 and 0.028.

5. V,K,J,X,Q,Z, each having probability less than 0.01.

It may also be useful to consider sequences of two or three consecutive letters called diagrams and trigrams respectively.

Common diagrams (in decreasing order): TH, HE, IN, ER, AN, RE, ED, ON, ES, ST, EN, AT, TO, NT, HA, ND, OU, EA, NG, AS, OR, TI, IS, ET, IT, AR, TE, SE, HI, OF.

Common trigrams (in decreasing order) are: THE, ING, AND, HER, ERE, ENT, THA, NTH, WAS, ETH, FOR, DTH.

Table 6.1: Probability of occurrence of 26 letters

letter	probability	letter	probability
A	.082	N	.067
B	.015	O	.075
C	.028	P	.019
D	.043	Q	.001
E	.127	R	.060
F	.022	S	.063
G	.020	T	.091
H	.061	U	.028
I	.070	V	.010
J	.002	W	.023
K	.008	X	.001
L	.040	Y	.020
M	.024	Z	.001

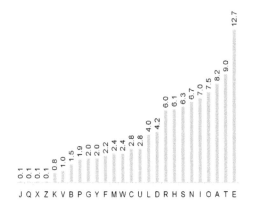

Figure 6.2: Statistical Distribution of English Alphabet

6.4.2   *Cryptanalysis of Affine Cipher*

Assume that an English plaintext is encrypted by an Affine cryptosystem (ignoring space and punctuation) as follows:

Example 97 *Consider the following ciphertext*

BHJUHNBULSVULRUSLYXHONUUNBWNUAXUSNLUYJ
SSWXRLKGNBONUUNBWSWXKXHKXDHUZDLKXBHJUHB
NUONUMHUGSWHUXMBXRWXKXLUXBHJUHCXKXAXKZ
SWKXXLKOLJKCXLCMXONUUBVULRRWHSHBHJUHNBX
MBXRWXKXNOZLJBXXHBNFUBHJUHLUSWXGLLKZLJPH
UULSYXBJKXSWHSSWXKXNBHBHJUHYXWNUGSWXGLLK

The frequency of 26 letters in the ciphertext in decreasing order is given in Table 6.2.

Now, we will compare the obtained frequency in Table 6.2 with the actual frequency in Table 6.1. We shall also make some guessing as follows:

First Guessing: Make the first guessing as $E = X, T = U$. We get the linear system of equations $4a + b = 23 \pmod{26}$ and $19a + b = 20 \mod 26$. Solving these two equations, we get $a = 5$ and $b = 4$. Using the secret key as $(5, 3)$ and decrypting the given ciphertext we get a text starting with "kgwtgcktmootmitdmzeg..", which is a meaningless text. So, our first guess is wrong. (See Table 6.3).

Second Guessing: In the second guessing, let, $E = X$ and $A = H$. We get the linear system of equations as $4a + b = 23 \pmod{26}$ and $b = 7 \pmod{26}$. On solving, we get : $a = 4$ or $a = 17$ and therefore $a = 17$ ($a = 4$ is not invertible). Thus, the shared secret key pair is $(17, 7)$. Using this secret key, we get the decryption Table 6.4 and the meaningful plaintext as below:

s a u n a i s n o t k n o w n t o b e a f i n n i s h i n v e n t i o n b u t t h e w o r d i s f i n n i s h t h e r e a r e m a n y m o r e s a u n a s i n f i n l a n d t h a n e l s e w h e r e o n e s a u n a p e r e v e r y t h r e e o r f o u r p e o p l e f i n n s k n o w w h a t a s a u n a i s e l s e w h e r e i f y o u s e e a s i g n s a u n a o n t h e d o o r y o u c a n n o t b e s u r e t h a t t h e r e i s a s a u n a b e h i n d t h e d o o r.

Shift cipher is a particular case of affine cipher ($a = 1$) here, we need to determine the value of b only. Using a similar technique, we can apply the frequency analysis on shift cipher also.

Table 6.2: Frequency of occurrence of 26 letters

y_i	f_i	y_i	f_i	y_i	f_i	y_i	f_i
X	32	S	15	Y	4	F	1
U	30	W	14	Z	4	P	1
H	23	J	11	C	3	E	0
B	19	O	6	A	2	I	0
L	19	R	6	D	2	O	0
N	16	G	5	V	2	T	0
K	15	M	4	V	2		

Table 6.3: Decryption in the First Guessing

Ciphertext	A	B	C	D	E	F	G	H	I	J	K	L	M	N
Plaintext	p	k	f	a	v	q	l	g	b	w	r	m	h	c
Ciphertext	O	P	Q	R	S	T	U	V	W	X	Y	Z		
Plaintext	x	s	n	i	o	y	t	o	j	e	z	u		

Table 6.4: Decryption in the Second Guessing

Ciphertext	A	B	C	D	E	F	G	H	I	J	K	L	M	N
Plaintext	v	s	p	m	j	g	d	a	x	u	r	o	l	i
Ciphertext	O	P	Q	R	S	T	U	V	W	X	Y	Z		
Plaintext	f	c	z	w	t	q	n	k	h	e	b	y		

6.4.3 Cryptanalysis of Substitution Cipher

We illustrate the concept of frequency analysis with the following example.

Example 98 *Consider a given ciphertext*

ODQSOCL OW GIU BOEE QRROHOCS QV GIUR KIA QF Q DQCQSLR WIR ICL IW CQFQF EIYQE YIDJUVLR FGFVLDF GIU SLV OCVI GIUR IWWOYL IC VXQV DICPQG DIRCOCS VI WOCP VXL JXICLF ROCSOCS LHLRG YQEELR OF Q POFVRQUSXV YICWUFLP CQFQ BIRMLR QCP LHLRG YQEELR QF-FURLF GIU VXQV XOF IR XLR WOEL IR QYYIUCVOCS RLYIRP IR RLFLQRYX JRIKLYV LHLRG ICL IW BXOYX OF DOFFOCS WRID VXL YIDJUVLR FGFVLD OF QAFIEUVLEG HOVQE.

Using frequency analysis, we will find the corresponding plaintext. We first obtain the frequency table of all 26 letters in the given ciphertext as given in Table 6.5. The second column represents frequency of alphabets, the third column represents the percentage of alphabets in the given ciphertext, and the fourth column is the associated plaintext alphabets. We get the associated plaintext alphabets by using finitely many guessings on ciphertext.

Table 6.5: Probability of occurrence of 26 letters

CT	Frequency	% in CT	PT	CT	Frequency	% in CT	PT
A	2	0.63	Q	N	0	0.00	C
B	3	0.95	A	O	23	7.28	I
C	21	6.65	Y	P	6	1.90	J
D	10	3.16	P	Q	25	7.91	N
E	12	3.80	L	R	29	9.18	R
F	23	7.28	W	S	10	3.16	F
G	12	3.80	S	T	0	0.00	V
H	5	1.58	X	U	12	3.80	U
I	31	9.81	O	V	20	6.33	H
J	4	1.27	K	W	10	3.16	B
K	2	0.63	M	X	11	3.48	T
L	31	9.81	E	Y	13	4.11	G
M	1	0.32	D	Z	0	0.00	Z

Using this correspondence between ciphertext and plaintext alphabets, we get a meaningful plaintext

IMAGINE IF YOU WILL ARRIVING AT YOUR JOB AS A MANAGER FOR ONE OF NASAS LOCAL COMPUTER SYSTEMS YOU GET INTO YOUR OFFICE ON THAT MONDAY MORNING TO FIND THE PHONES RINGING EVERY CALLER IS A DISTRAUGHT CONFUSED NASA WORKER AND EVERY CALLER ASSURES YOU THAT HIS OR HER FILE OR ACCOUNTING RECORD OR RESEARCH PROJECT EVERY ONE OF WHICH IS MISSING FROM THE COMPUTER SYSTEM IS ABSOLUTELY VITAL.

6.4.4 Cryptanalysis of Vigenere Cipher

In poly-alphabetic cipher, one plaintext character is replaced by more than one ciphertext characters, so the letter frequencies are obscured but not totally lost. At a first glance, the poly-alphabetic cipher seems to be intractable against frequency analysis. In this section, we discuss that frequency analysis is also possible in case of poly-alphabetic cipher, in particular in Vigener cipher. Vigenere cipher can be visualized as an application of m shift cipher, if the key length is m. For cryptanalysis of Vigenere cipher, we divide the whole ciphertext into m classes of ciphertexts where each class is encrypted by a single shift cipher. So, if the length of keyword is known, then by repeated application of frequency analysis we can obtain the actual keyword. First we will discuss how to find the key length and then how to find the actual keyword.

For Vigenere cipher, as the length of the keyword increases, the letter frequency shows less English-like characteristics and becomes more random. There are Two methods to find the key length: (1) Kasiski test and (2) Index of coincidence method.

Kasiski Test: This test was proposed by Friedrich Kasiski in 1863. The idea behind Kasiski test is that two identical segments of plaintext will be encrypted to the same ciphertext whenever their occurrence in the plaintext is δ positions apart, where $\delta \equiv 0 \pmod{m}$. And, conversely, if we observe two identical segments of ciphertext, each of length at least three, say, then there is a good chance that they correspond to identical segments of plaintext. To find the key length using Kasiski test, we search for pairs of identical segments of length at least three. Then we record the distances between starting position of two segments: $\delta_1, \delta_2 \ldots$ etc. We would conjecture that m divides all of δ_is, and hence m divides $\gcd(\delta_1, \delta_2, \ldots)$. Consider the following example.

Example 99 *Let the secret key of Vigenere cipher be "KING" and the plaintext be "the sun and the man in the moon". Using the encryption key "KING", we obtain the ciphertext "D P R Y E V N T N $\underline{B U K}$ W I A O X B $\underline{U K}$ W W B T".*

If we observe the plaintext-ciphertext pair carefully, we see that the segments "the" appears in the 10th and 18th position in the plaintext and is encrypted to the same ciphertext "BUK". The distance between two "BUK" is 8. It is very likely that the key length will be either 2, 4 or 8.

Index of Coincidence: Using the index of coincidence, we not only find the key length but also the actual key. This method was proposed by William F. Friedman in 1920. Index of Coincidence (IC) is a statistical measure which distinguishes the original text with the text encrypted with a substitution cipher.

Definition 6.2 *Index of Coincidence*: Suppose $x = x_1x_2x_3...x_n$ is a string of n alphabet characters. The index of coincidence of x, denoted by $I_c(x)$ is defined to be the probability that two random elements of x are identical.

Thus $I_c(x)$ is defined to be the probability that two randomly selected letters of x will be identical. Suppose $f_0, f_1, f_2...f_{25}$ denote the frequencies of English alphabet $A, B, C, ...Z$ in x. We can choose any two elements in x by ${}^nC_2$ ways. For $0 \leq i \leq 25$, there are ${}^{f_i}C_2$ ways choosing both elements to be i. Thus, the formula for index of coincidence is

$$I_c(x) = \frac{\sum_{i=0}^{25} {}^{f_i}C_2}{{}^nC_2} = \frac{f_i(f_i - 1)}{n(n-1)} \tag{6.1}$$

Now, suppose x is a string of English language. Let $p_0, p_1, p_2, ..., p_{25}$ denote the expected probability of occurrence of $A, B, C, ...Z$ in the string x (as given in the Table 6.1). Since the probability, that two random elements; both are A is p_0^2, both are B is p_1^2 etc., then we would except that

$$I_c(x) \approx \sum_{i=0}^{25} p_i^2 \approx 0.065 \tag{6.2}$$

If x is a ciphertext obtained by using any mono-alphabet cipher, then the individual probability of alphabet will be permuted, but the sum in equation (6.2) remains same.

We can visualize the equation (6.2) as follows. Imagine a hat is filled with 26 letters of the English alphabet uniformly. The chance of pulling out an A is 1/26. If we had two such hats, the probability of pulling out two A's simultaneously is $(1/26) \times (1/26)$. The chance of drawing any pair of letters is $26 \times (1/26) \times (1/26) = (1/26) = 0.0385$. Now, suppose we fill the hats with 1000 letters ($n = 1000$, here) and had the number of each letter correspond to the average frequency of that letter in the English language (i.e. 82 A's, 15 B's, 28 C's, etc.). The chance of drawing any pair of identical letters is $(82/1000)(82/1000) + (15/1000)(15/100) + (28/100)(28/100) + ... = 0.0.065$. Thus, the index of coincidence of English language $I_C(English) = 0.065$.

Example 100 *Consider a text "x = THE INDEX OF COINCIDENCE". Counting the frequencies of alphabet, we see that* $f_0 = 0, f_1 = 0, f_2 = 3, f_3 = 2, f_4 = 4, f_5 = 1, f_6 = 0, f_7 = 1, f_8 = 3, f_9 = f_{10} = f_{11} = f_{12} = 0, f_{13} = 3, f_{14} = 2, f_{15} = f_{16} = f_{17} = f_{18} = 0, f_{19} = 2, f_{20} = f_{21} = f_{22} = 0, f_{23} = 1, f_{24} = f_{25} = 0.$ *Thus, the index of coincidence of the text x is given by*

$$I_c(x) = \sum_{i=0}^{i=25} \frac{f_i(f_i - 1)}{n(n-1)} = \frac{34}{420} = 0.0809$$

Key Length: Suppose, we have given a ciphertext string $y = y_1 y_2 y_3 .... y_n$ that has been constructed using a Vigenere cipher. We want to determine the key length. We divide the ciphertext string y, into m substrings denoted by $u_1, u_2, u_3, ... u_m$ respectively. Each u_i is called a coset of y, where

$$
\begin{aligned}
u_1 &= y_1 y_{m+1} y_{2m+1} \cdots \\
u_2 &= y_2 y_{m+2} y_{2m+2} \cdots \\
u_3 &= y_3 y_{m+3} y_{2m+3} \cdots \\
.. &= \ldots\ldots\ldots \\
u_m &= y_m y_{2m} \cdots
\end{aligned}
$$

If $u_1, u_2, ... u_m$ are constructed in this way and m is the actual keyword length, then each u_i are obtained using a single mono-alphabetic cipher. Hence, the index of coincidence of each u_i will be roughly equal to 0.065, and the average of these IC's would still be high and close to $I_c(English) = 0.065$. Otherwise, alphabets of each u_i are much randomly distributed, as each u_i are obtained by shift encryption with different keys. Hence, index of coincidence $I_c(u_i)$ for each $i = 1, 2, 3...m$, becomes more close to 0.038. Hence, the average of the IC's would be low. Based on this observation, we may divide the ciphertext into 1 coset (the ciphertext itself), 2 cosets, 3 cosets, 4 cosets, etc. and compute the IC of each coset and the average. The length that yields the highest average IC and close to $I_c(English)$ is likely to be the correct length of the keyword. In this way, we can easily determine the length of the keyword. Let us see the following example.

Example 101 *Suppose we have the ciphertext "**dlc** srbob mp gmsrashcxgc sw y wiycypo sd rsu cmkspyb e dbioeilmc bswrbmzexgyr gc xm **dlc** ergpspw hgcxpsfsdmmx". If we apply the Kasiski test, the triagram dlc will appear in the ciphertext in the 1st and 70th position. The difference between this is 69. Thus, the key length m divides 69, which shows that the key length is either 3 or 23. To find key length using IC, we first find the frequency count of CT as given in Table 6.6.*

The Table 6.7 shows a summary of possible keyword lengths from $m = 1$ *to* $m = 4$*.*

From the above table, we see that the maximum average of index of coincidence is 0.0656 *for* $m = 3$*, which is very close to* $I_c(Engilsh) = 0.065$*. Thus, the key length must be 3.*

Table 6.6: Frequency Count in CT

x_i	A	B	C	D	E	F	G	H	I	J	K	L	M
f_i	1	6	9	5	4	1	6	2	3	0	1	3	8
x_i	N	O	P	Q	R	S	T	U	V	W	X	Y	Z
f_i	0	3	6	0	6	1	0	1	0	4	5	5	1

Table 6.7: $I_c(u_i)$ to find keyword length

Length (m)	Index of Coincidences (IC)	IC(Average)
1	0.0564	0.0564
2	0.0536, 0.0525	0.0535
3	0.086, 0.051, 0.059	0.0656
4	0.0711, 0.0672, 0.0553, 0.0432	0.0592

The Actual Keyword: Suppose, we have determined the keyword length (m, say). Now we want to determine the actual key $K = (k_1, k_2, k_3 ... k_m)$. Let $1 \leq i \leq m$, and let $f_0, f_1, f_2, ... f_{25}$ denote the frequency of alphabets $A, B, C, D, ...Z$ respectively in the string u_i. Also, assume that $n' = \frac{n}{m}$ denotes the length of u_i. Then the probability distribution of alphabets (26 letters) in u_i is $\frac{f_0}{n'}, \frac{f_1}{n'}, ... \frac{f_{25}}{n'}$. Clearly, the substring u_i of the ciphertext y is obtained by shift encryption of a part of plaintext elements using a single key k_i. Therefore, we would expect that the shifted probability distribution $\frac{f_{k_i}}{n'}, \frac{f_{k_i+1 \ (mod \ 26)}}{n'}, ... \frac{f_{k_i+25 \ (mod \ 26)}}{n'}$ will be close to the ideal probability distribution $p_0, p_1, p_2, ... p_{25}$. For any $0 \leq g \leq 25$, we find the mutual index of coincidence M_g defined as

$$M_g = \sum_{i=0}^{i=25} p_i \frac{f_{i+g \ (mod \ 26)}}{n'} \tag{6.3}$$

Now if $g = k_i$, then we would expect that $M_g \approx \sum_{i=0}^{i=25} p_i^2 = 0.065$. Thus, for each substring u_i, we determine M_g. If for any $0 \leq g \leq 25$, the $M_g \approx 0.065$, then we expect that $g = k_i$. Let us find the actual key of the previous example i.e. Example 101.

Example 101(Cont.) Since the keyword length is 3, we have the following three cosets of PT.

$$u_1 = dsopssxswcorcsbbemsbeycdepwcsdx$$

$$u_2 = lrbgrhgwiyssmpeiicwmxrxlrshxfm$$

$$u_3 = cbmmaccyypdukydolbrzggmcgpgpsm$$

Now we find the mutual index of coincidence for each u_i as given in Table 6.8. From Table 6.8, we see that the key is likely to be $K = (10, 04, 24)$, and hence, the keyword is "KEY".

Table 6.8: Mutual Index of Coincidence M_g calculation

	Value of $M_g(u_i)$						
i=1	0.047	0.043	0.025	0.030	0.046	0.034	0.024
	0.028	0.033	0.038	0.062	0.055	0.030	0.029
	0.060	0.045	0.037	0.027	0.039	0.029	0.034
	0.032	0.040	0.041	0.044	0.050		
i=2	0.041	0.034	0.027	0.042	0.064	0.042	0.031
	0.030	0.040	0.037	0.039	0.036	0.031	0.040
	0.046	0.045	0.033	0.035	0.038	0.049	0.043
	0.029	0.025	0.037	0.047	0.041		
i=3	0.029	0.034	0.046	0.028	0.032	0.029	0.040
	0.037	0.045	0.036	0.048	0.045	0.037	0.038
	0.041	0.038	0.029	0.031	0.034	0.039	0.047
	0.044	0.034	0.034	0.064	0.045		

Using the keyword "KEY" decrypting the ciphertext, we find the plaintext as follows.

"The index of coincidence is a measure of how similar a frequency distribution is to the uniform distribution."

In this example, number of characters $n = 91$ and the $I_c(PT) = 0.065$ but the $I_c(CT) = 0.0564$. $I_c(CT)$ is more close to 0.065, which implies that ciphertext characters are less randomized. This shows that the key length is small.

6.4.5 Cryptanalysis of Hill Cipher

It is difficult to break the Hill cipher using ciphertext only attack but it is not secure against known plaintext attack. Assume that the attacker has determined the order m of the key matrix being used in the Hill cipher. Also, suppose that the attacker has at least m distinct plaintext-ciphertext pairs, say $\{(x_j, y_j) : y_j = e_K(x_j), \ for \ 1 \le j \le m \ and \ x_j = (x_{1,j}, x_{2,j}, ...x_{m,j}), \ y_j = (y_{1,j}, y_{2,j}, ..., y_{m,j})\}$. By the encryption of the Hill cipher, we have $x_j = y_j \times K$, for $1 \le j \le m$. If we define $X = (x_{i,j})_{m \times m}$ and $Y = (y_{i,j})_{m \times m}$, then we have the matrix equation as $Y = XK$. If the matrix X is invertible, then we can determine the encryption key $K = X^{-1}Y$ and hence can break the system. (If the matrix X is not invertible, then we try to apply the same process for the other sets of m plaintext-ciphertext pairs). Let us do the following toy example.

Example 102 *Suppose the ciphertext is "ENDQRH", which is encrypted using a key of order 2×2 matrix. Let the attacker know two plaintext-ciphertext pairs as (nd, DQ) and (on, RH) i.e. $e_K(13, 3) = (3, 16)$ and $e_K(14, 13) = (17, 7)$. Thus, $x_1 = (13, 3), x_2 = (14, 13)$ and $y_1 = (3, 16), y_2 = (17, 7)$.*

Thus, the matrix equation is $Y = XK$, where $X = \begin{pmatrix} 13 & 3 \\ 14 & 13 \end{pmatrix}$ and $Y =$
$\begin{pmatrix} 3 & 16 \\ 17 & 7 \end{pmatrix}$*. Since $|X| \pmod{26} = 23 \Rightarrow X$ is invertible and $X^{-1} = \begin{pmatrix} 13 & 1 \\ 22 & 13 \end{pmatrix}$.*
Thus, the encryption key K can be computed as

$$K = X^{-1}Y = \begin{pmatrix} 13 & 1 \\ 22 & 13 \end{pmatrix} \begin{pmatrix} 3 & 16 \\ 17 & 7 \end{pmatrix} = \begin{pmatrix} 4 & 7 \\ 1 & 1 \end{pmatrix}$$

Since $|K| \bmod 26 = 23 \neq 0$. The matrix K is invertible. We have

$$K^{-1} = \begin{pmatrix} 17 & 11 \\ 9 & 16 \end{pmatrix}.$$

Decrypting the given ciphertext message "ENDQRH" using the key K^{-1}, we get the original message as "london".

Even if the attacker does not know the order m of key matrix, then assuming that the order m is not too big, he could simply try for $m = 2, 3, 4...$ etc. In this way, the obtained key matrix can be applied to further plaintext-ciphertext pair. If the key matrix is not correctly calculated, then it will not satisfy this pair of plaintext and ciphertext.

6.4.6 Cryptanalysis of LFSR Stream Cipher

The LFSR is also vulnerable to known-plaintext attack like Hill cipher.

Recall that in LFSR, we generate a keystring bits $(z_1, z_2, z_3....)$ of higher period using an initial keystring $(k_1, k_2, ...k_m)$ and a linear recurrence relation defined as follows

$$z_{i+m} = \sum_{j=0}^{m-1} c_j z_{i+j} \bmod 2, \quad \forall i \geq 1, \quad \text{where} \quad z_i = k_i, i = 1, 2, 3...m \text{ and}$$
$c_0, c_1, c_2, ... \in \mathbb{Z}_2$ are predefined constants.

The encryption of plaintext $(x_1, x_2, x_3...)$ is just addition modulo 2 operation with the key string. So, to determine the whole plaintext from the given ciphertext, the attacker needs to obtain the linear recurrenc relation i.e. the values $(c_0, c_1, c_2, ...c_{m-1})$ and the initial keyword $(k_1, k_2, k_3...k_m)$. Assume that the attacker knows the length of initial keyword m. Suppose the attacker has a plaintext string $(x_1, x_2, x_3, ...x_n)$ and the corresponding ciphertext string $(y_1, y_2, y_3, ...y_n)$ where $n \geq 2m$. The attacker can compute $z_i = (x_i + y_i) \bmod 2$ for $1 \leq i \leq n$. To determine the complete key string bits, he needs to determine the m unknown $c_0, c_1, c_2, ...c_{m-1}$. Using the recurrence relation, we can obtain the system of matrix equation

$$\begin{pmatrix} z_{m+1} & z_{m+2} & \cdot & \cdot & \cdot & z_{2m} \end{pmatrix} = \begin{pmatrix} c_0 & c_1 & \cdot & \cdot & \cdot & c_{m-1} \end{pmatrix} K$$

Where $K = \begin{pmatrix} z_1 & z_2 & \cdot & \cdot & \cdot & z_m \\ z_2 & z_3 & \cdot & \cdot & \cdot & z_{m+1} \\ \cdot & \cdot & & & & \cdot \\ \cdot & \cdot & & & & \cdot \\ \cdot & \cdot & & & & \cdot \\ z_m & z_{m+1} \cdot & \cdot & \cdot & \cdot & z_{2m} \end{pmatrix}.$

If the coefficient matrix K is invertible (modulo 2), then the solution is
$\begin{pmatrix} c_0 & c_1 & \cdot & \cdot & \cdot & c_{m-1} \end{pmatrix} = \begin{pmatrix} z_{m+1} & z_{m+2} & \cdot & \cdot & \cdot & z_{2m} \end{pmatrix} K^{-1}.$
Consider the following example.

Example 103 *Let the attacker knows the following pair of ciphertext and plaintext string:*

$PT = 1001\ 0010\ 0110\ 1101\ 1001\ 0010\ 0110,$
$CT = 1011\ 1100\ 0011\ 0001\ 0010\ 1011\ 0001.$

On XORing, we get keystream as 0010 1110 0101 1100 1011 1001 0111. *Observing the keystream carefully, we can group them in blocks of 7 bits as* 0010111 0010111 0010111 0010111.

By Theorem 6.2 "The maximum sequence length generated by an LSFR of degree m is $2^m - 1$", we can guess that the degree might be 3 as the sequence length of the LSFR appears to be 7. Thus, it is a 3-stage LFSR i.e. $m = 3$. Clearly, $z_1 = 0, z_2 = 0, z_3 = 1$.
To determine the constants c_0, c_1, c_2 of recurrence relation $z_{i+3} = \sum_{j=0}^{2} c_j z_{i+j}$ mod 2, we need to solve the system of equation.

$$(0, 1, 1) = (c_0, c_1, c_2) \begin{pmatrix} 0 & 0 & 1 \\ 0 & 1 & 0 \\ 1 & 0 & 1 \end{pmatrix}$$

It can be easily verified that

$$\begin{pmatrix} 0 & 0 & 1 \\ 0 & 1 & 0 \\ 1 & 0 & 1 \end{pmatrix}^{-1} = \begin{pmatrix} 1 & 0 & 1 \\ 0 & 1 & 0 \\ 1 & 0 & 0 \end{pmatrix}$$

Thus, $(c_0, c_1, c_2) = (0, 1, 1) \begin{pmatrix} 1 & 0 & 1 \\ 0 & 1 & 0 \\ 1 & 0 & 0 \end{pmatrix} = (1, 1, 0).$

Thus, the recurrence used to generate the keystream is
$z_{i+3} = (z_i + z_{i+1})$ mod 2.

6.5 Exercises

1. In Affine cipher, if the encryption is $E_k(x) = 3x + 1 \mod 26$, and the ciphertext is E, then determine the plaintext.

2. Consider the Shift cipher and assume that the plaintext message is in English alphabet. Decrypt the following ciphertext, where ciphertext is written in five letter groups, using the frequency analysis with brief explanation.
EAIPP ORSAR IBEQT PISJE WYFWX MXYXM SRGMT LIVMW XLIGE IWEVG MTLIV XSIRG VCTXE QIWWE KIAMX LXLIG EIWEV GMTLI VIEGL PIXXI VSJQI WWEKI MWVIT PEGIH FCXLI PIXXI VXLVI ITSWM XMSRW PEXIV MRXLI EPTLE FIX

3. Consider the Affine cipher by selecting English alphabet (ABCDEFGHIJKLMNOPQR
STUVWXYZ). Using frequency analysis, decrypt the following ciphertext, where ciphertext is written in five letter groups.
IJCEE AICJI AVXOB ECBOI SMMSP JWYKA HOOCE WHSFB OCGMC PWSRH XOIJC EEAIC JIAVX OBEIC PFOFB SGOPO NOPAR HXOCH HCIGO BSPJW GPSQE EKRRA IAOPH IAVXO BHOTH CPLXO PIOHX OWCBO EKEIO VHAFJ OHSCI AVXOB HOTHS PJWCH HCIGE SMOIJ CEEAI CJIAV XOBEX CNOCE MCJJG OWEVC IOHXO EOIAV XOBEI CPFOF BSGOP QAHXC FBKHO RSBIO CHHCI GHXCH AEFWE AMVJW HBWAP USKHC JJGOW EEKFE HAHKH ASPIA VXOBE ICPXC NOCJC BUOGO WEVCI OFKHC BOSRH OPEKE IOVHA FJOHS CRBOY KOPIW CPCJW EAEFO ICKEO RSBOT CMVJO RBOYK OPHJO HHOBE APHXO VJCAP HOTHJ CPUKC UOISB BOEVS PLHSR BOYKO PHJOH HOBEA PHXOI AVXOB HOTHE

4. The ciphertext SDJ was encrypted using the Affine cipher $7x + 3 \mod 26$. Find the plaintext.

5. The following ciphertext was encrypted by an Affine cipher $\mod 26$: DJHGDKA. If the first two alphabets are *am*, then find the ciphertext and the secret key (a, b).

6. Suppose we encrypt any plaintext using an Affine cipher, and then again encrypt the resultant ciphertext using another Affine cipher under modulo 26. Does this increases the security? Justify your answer.

7. A plaintext is encryption by "OSHX OS DX OWGRKUKX DSKMPR WLG GKSR" using Vigenere cipher with key length four. If the attacker knows the first and the last four alphabet of plaintext as *meet* and *shop*. Then, find the complete plaintext.

8. The following ciphertext were obtained by using Vigenere cipher maintaining the original punctuation.
EENX CAOI. V CTLM WNOE YPU OKSIOD GNP OMD OATLEIAM

ZF SAWG CAOJRKE SJNTN. CENEZHPR UHR SPEUIAM EINE VY EWFLIK AM. UHVY TS PUE RLSU CUGYCF TB SPEU. ASZPR UHVY XEFTVTR I XIYR ME PUG UQ SUAGOZN GOE UYE NOAZS. WIEA O HIML PUXE CAPQ HE XIYR REU MNXCIFD. XKPP UHVY TN-GOESLTJOA YPCSEG, ZZ NPT GKWL UO NTJBPDL KGEO YBAC CMOFK QRJEAJ LLTO. VL JOVR SGEHFR JOWL LNBC LBPUG UFR NERZTNH, HR SLY HEG GYGSY.

9. The ciphertext WZYFZUCWUXOS was produced by a Hill cipher (using encryption rule $MP = C$, where P is the plaintext, M is the matrix key, and C is the ciphertext) with matrix $\begin{pmatrix} 13 & 12 \\ 22 & 5 \end{pmatrix}$. Find the plaintext.

10. The ciphertext KOTJPQHWLE was produced by a Hill cipher using a 2×2 matrix key. If the plaintext is *singapore*, find the encryption matrix.

11. Determine the number of different keys in Affine cipher if the alphabet is alphanumeric $(a, b, c, d, \ldots 0, 1, 2, 3 \ldots 9)$. Describe the encryption and decryption process in this case.

Chapter 7

Block Ciphers

CONTENTS

7.1   Introduction

Block ciphers are one of the most important tools in modern cryptography. They are used not only for encryption but also for construction of stream ciphers, hash functions, and many other primitives.

While stream ciphers encrypt individual characters of plaintext message one at a time using an encryption rule which may vary with plaintext, time, and state, block ciphers process plaintext in relatively large blocks using the same function to encrypt successive blocks. For example, considering each letter of the English alphabet as a plaintext character, the substitution cipher and the Vigenère cipher, introduced in Chapter 6 are stream ciphers, while the Hill cipher and the permutation cipher are block ciphers. Note that, if we consider the letters of the English alphabet as 8-bit binary strings, then all of these ciphers are block ciphers. Further, stream ciphers have 'memory' while block ciphers are memoryless and one can obtain a stream cipher by adding suitable 'memory' to a block cipher. One common property between most of the block ciphers and stream ciphers is that they are not unconditionally secure and only attempt to achieve computational security.

Remark 7.1   Most public-key cryptosystems, such as RSA and ElGamal, introduced in Chapter 9 are block ciphers but in this chapter (and almost everywhere in this book) we will use the term block cipher only for symmetric block ciphers, unless otherwise stated.

Most of today's block ciphers constructions use the basic design principles introduced by Shannon [154] in 1948. In fact, many of the block ciphers in use today, with some notable exceptions, draw heavily from the two most popular block ciphers DES and AES. DES [122] was designed at IBM under the supervision of Feistel and was designated the US government FIPS standard [122] for protecting sensitive but unclassified data in 1977. And AES [51] was the winner of the NIST's open AES competition [123] and was designated the US government FIPS standard [124] in 2000. In this chapter, we will discuss the basic design considerations while constructing a block cipher in Section 7.4, some modes of operations of block ciphers in Section 7.2, and introduce the DES and AES algorithms in Sections 7.5 and 7.6 respectively.

7.1.1 Notations

Block ciphers process plaintext in relatively large blocks of a fixed length, n, called the *block size*, to output ciphertext of a fixed length m. Generally, both the *input plaintext* and the *output ciphertext* are of the same length, n, so that the set of all possible input plaintexts, $\mathcal{P}$, and the set of all possible output ciphertexts, $\mathcal{C}$, are both $\{0,1\}^n$. The *encryption key*, K, which is also the *decryption key*, is also of a fixed length, k, called the *key size*. The key size k may or may not be the same as the block size n. We denote the set of all possible keys by $\mathcal{K} = \{0,1\}^k$. The encryption and the decryption algorithms of a given block cipher are denoted by E and D, and E_K and D_K denote these operations with K as the input key.

Note that, for each $K \in \mathcal{K}$, the E_K must be injective to allow an unambiguous decryption. Thus, m must be greater than or equal to n. When $m > n$, data expansion occurs which is not desirable. When $m = n$, $\mathcal{P} = \mathcal{C} = \{0,1\}^n$, E_K is a bijection on $\{0,1\}^n$, and $D_K = E_K^{-1}$. So, we have the following mathematical definition of block ciphers:

Definition 7.1   An n-bit *block cipher* is a pair of functions $E : \mathcal{P} \times \mathcal{K} \to \mathcal{C}$ and $D : \mathcal{C} \times \mathcal{K} \to \mathcal{P}$, where $\mathcal{P} = \{0,1\}^n$, $\mathcal{C} = \{0,1\}^m$, and $\mathcal{K} = \{0,1\}^k$, and $m \geq n$, such that, for all $K \in \mathcal{K}$ and for all $P \in \mathcal{P}$,

$$D_K(E_K(P)) := D(E(P,K),K) = P.$$

Remark 7.2   Ideally, each $K \in \mathcal{K}$, should define a distinct E_K.

Remark 7.3   For the block ciphers considered in this chapter (and throughout this book), we will assume $m = n$ and $\mathcal{P} = \mathcal{C} = \{0,1\}^n$, unless otherwise stated.

If a plaintext is shorter than the block size, we use *padding* (see Subsection 7.3) to process the plaintext so that the input to the block cipher is of length n. If the plaintext is longer than the block size, the plaintext is divided into blocks of length n, (if necessary, padding is done to ensure that the plaintext is of a length which is a multiple of n) and then encrypted using a suitable *mode of operation*.

7.2   Modes of Operation

Block ciphers are designed to encrypt plaintext blocks of fixed length, n, generally a power of 2, such as 8, 16, 32, 64, 128, and 256. Many methods have been proposed for using a block cipher to encrypt long messages. Such a method is referred to as a *mode of operation*. The modes of operation are designed to achieve efficiency, security, ease of use, many other functionalities and their trade-off as per the requirements, and the block cipher in use. We will discuss these properties with respect to the five most common modes of operation:

- ■ Electronic Code Book (ECB)

- ■ Cipher-Block Chaining (CBC)

- ■ Cipher FeedBack (CFB)

- ■ Output FeedBack (OFB)

- ■ Counter (CTR)

We will assume that the plaintext x is of a length which is a multiple of the block size, say ℓn, if necessary by applying a suitable padding (see Subsection 7.3). The plaintext is divided into ℓ blocks, $x_1, x_2, \ldots, x_\ell$, each of length n. These, then, become inputs to the chosen mode of operation which outputs ℓ ciphertext blocks, $c_1, c_2, \ldots, c_\ell$, each of length n.

Some modes also require some more auxiliary information as input such as an *initialization vector* (IV) or a shift register, which may be an input by the user or the mode itself may generate it. The IV is generally a *nonce* used in some modes of operation to introduce some randomness to the otherwise deterministic block cipher. The IV ensures that identical plaintexts produce distinct ciphertexts. The IV input to the mode is denoted by x_0 and the corresponding output is denoted by c_0, often, $c_0 = x_0$. The *shift register r* is an integer value between 1 and n and is used in some modes of operation to enable encryption of r-bit blocks. When r is 1, the system acts as a stream cipher.

7.2.1 Electronic Code Book (ECB)

The electronic code book (ECB) mode of operation is the simplest approach for encrypting messages longer than the block size. It simply divides the plaintext x into ℓ n-bit blocks, $x_1, x_2, \ldots, x_\ell$, and encrypts each of the blocks independently. Similarly, each of the ciphertext blocks, $c_1, c_2, \ldots, c_\ell$, are decrypted independently

- ■ **Encryption:**
  $c_i \leftarrow E_K(x_i)$, for all $i = 1, \ldots, \ell$.

- ■ **Decryption:**
  $x_i \leftarrow D_K(c_i)$, for all $i = 1, \ldots, \ell$.

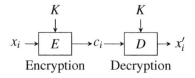

Figure 7.1: Electronic Code Book Mode

7.2.1.1 ECB Properties

The ECB mode of operation is the simplest to implement. Each block is encrypted independently which not only allows for parallellized implementation in hardware and software but also allows for random access which is important for efficient hard disk encryption. Further, since all the blocks are encrypted independently, corruption of any of the ciphertext blocks affects the decryption of only that block and not of any of the other blocks.

But, despite its simplicity and efficiency, the ECB mode is not recommended for encrypting long messages or if keys are reused for encryption. This is due to the fact that identical plaintexts are encrypted to identical ciphertexts which can leak the information whether the same plaintext was encrypted or not and thus can leak data patterns. For example, see the image encrypted using ECB mode in Figure 7.2. Thus, modes other than ECB, which result in pseudo-random ciphertexts are used to encrypt long messages.

Original image Encrypted using ECB mode Encrypted using other modes

Figure 7.2: Drawback of the ECB mode

Let us do the following example.

Example 104 *Consider the block length $n = 4$ and the key space $\mathcal{K} = S_4$. Let the encryption function $E_\pi : \{0,1\}^4 \rightarrow \{0,1\}^4$ be defined by $E_\pi(b_1 b_2 b_3 b_4) = b_{\pi(1)} b_{\pi(2)} b_{\pi(3)} b_{\pi(4)}$ for some permutation key $\pi \in S_4$. Let the key π is defined by $\pi(1) = 4, \pi(2) = 1, \pi(3) = 2, \pi(4) = 3$.*

Let the plaintext be $m = 10100010100010111000$. Divide the whole message into fixed 4 bits of plaintext block as $x_1 = 1010, x_2 = 0010, x_3 = 1000, x_4 = 1011, x_5 = 1000$. Encrypting block-wise we get the ciphertext blocks as $c_1 = 0101, c_2 = 0001, c_3 = 0100, c_4 = 1101, c_5 = 0100$. Thus, the resultant ciphertext is $c = 01010001010011010100$.

7.2.2 Cipher-Block Chaining (CBC)

The cipher-block chaining (CBC) mode of operation is one of the oldest approaches which outputs pseudo-random ciphertexts. An *initialization vector* (IV), which is generally a freshly chosen random *nonce*, is used in the first block to ensure that each message is distinct and introduces an unpredictability to the otherwise deterministic block cipher. This initialization vector should be known by the sender and the receiver. In this mode, each block of the plaintext is XORed with the previous ciphertext block before being encrypted so that each ciphertext block depends on all the plaintext blocks processed up to that point ensuring that the effect of the IV propagates through all the blocks.

■ **Encryption:**
   $c_0 \leftarrow IV; c_i \leftarrow E_K(c_{i-1} \oplus x_i)$

■ **Decryption:**
   $c_0 \leftarrow IV; x_i \leftarrow c_{i-1} \oplus D_K(c_i)$

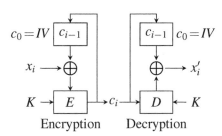

Figure 7.3: Cipher-Block Chaining Mode

7.2.2.1 CBC Properties

Note that in the CBC mode, even a one-bit change in the IV (or a plaintext block) affects all the following ciphertext blocks. Thus, even if identical plaintexts are encrypted with the same key, by choosing a fresh IV for every instance, the CBC mode outputs different ciphertexts and thus hides data patterns and avoids the information leakage which plagues the ECB mode.

But this gain in security comes at the cost of efficiency. The *chaining dependency* during encryption causes the loss of the parallelizability and the random access property during encryption. The decryption does not suffer though. Since the ith plaintext block $x_i = c_{i-1} \oplus D_K(c_i)$, the decryption process can be parallelized and the ith block of plaintext can be recovered independently from the blocks as long as the ith and the $(i-1)$th ciphertext blocks are available.

Since a one-bit change in the IV (or a plaintext block) affects all the following ciphertext blocks, if the plaintext needs any modification in the ith block, not only the *i*-th block but all the subsequent blocks must also be re-encrypted. But fortunately, *error propagation* is not so catastrophic during decryption. As mentioned earlier, the correct recovery of the *i*th plaintext block x_i requires the correctness of only the *i*th and the $(i-1)$th ciphertext blocks. A *r*-bit change/error in a ciphertext block causes a complete corruption of the corresponding block of plaintext and inverts the corresponding bits in the following block of plaintext, but the rest of the blocks remain intact. Thus, the CBC mode has the ability to *self-synchronize* and recover from errors. Consider the following example.

Example 105 *Consider the block length* $n = 4$ *and the key space* $\mathcal{K} = S_4$. *Let the encryption function* $E_\pi : \{0,1\}^4 \to \{0,1\}^4$ *be defined by* $E_\pi(b_1 b_2 b_3 b_4) = b_{\pi(1)} b_{\pi(2)} b_{\pi(3)} b_{\pi(4)}$ *for some permutation key* $\pi \in S_4$. *Le the key* π *is defined by* $\pi(1) = 4, \pi(2) = 1, \pi(3) = 2, \pi(4) = 3$.
  Let the initial vector $IV = 1111$ *and the plaintext be* $m = 10100010100010111000$. *Divide the whole message into fixed* 4 *bits of plaintext block as* $x_1 = 1010, x_2 = 0010, x_3 = 1000, x_4 = 1011, x_5 = 1000$. *Encryption of m using CBC mode is given in the following table.*

j	x_j	c_{j-1}	$c_{j-1} \oplus x_j$	$c_j = E_\pi\{c_{j-1} \oplus x_j\}$
1	*1010*	*1111*	*0101*	*1010*
2	*0010*	*1010*	*1000*	*0100*
3	*1000*	*0100*	*1100*	*0110*
4	*1011*	*0110*	*1101*	*1110*

Thus, the resultant ciphertext is 1010010001101110.

7.2.3 *Cipher FeedBack (CFB)*

The cipher feedback (CFB) mode of operation is a close relative of the CBC mode and shares many properties of the CBC mode. It also makes use of an IV and block chaining to output distinct ciphertexts at each instance. In CFB mode, the encryption function is not used directly for encrypting plaintext block but for generating a sequence of key blocks. The plaintext is encrypted by adding the key blocks mod 2. The ciphertext is decrypted by adding the same key block mod 2. The encryption is almost identical to that of the CBC mode—just the order of the XOR and the block cipher encryption is reversed—the encryption of each block of the plaintext is XORed with the previous ciphertext block. Because of the symmetry of the XOR operation, encryption and decryption are exactly the same. The key blocks can be simultaneously generated by the sender, Alice, and the receiver, Bob. Only addition mod 2 must be done sequentially.

■ **Encryption:**
  $c_0 \leftarrow IV; c_i \leftarrow E_K(c_{i-1}) \oplus x_i$

■ **Decryption:**
  $c_0 \leftarrow IV; x_i \leftarrow E_K(c_{i-1}) \oplus c_i$

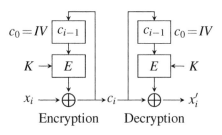

Figure 7.4: Cipher FeedBack Mode

7.2.3.1 CFB Properties

As in the CBC mode, the CFB mode also has the *chaining dependency* during encryption which causes the loss of the parallelizability and the random access property during encryption. But the decryption does not suffer since the ith plaintext block $x_i = c_i \oplus E_K(c_{i-1})$ can be recovered independently of other blocks as long as the ith and the $(i-1)th$ ciphertext blocks are available. Similarly, during decryption, the *error propagation* is limited and the decryption process *self-synchronizes* to recover from errors—though an r-bit change/error in a ciphertext block causes a complete corruption of the corresponding block of plaintext and inverts the corresponding bits in the following block of plaintext, but the rest of the blocks remain intact.

One of the main differences between the CFB mode and the CBC mode is that the block cipher decryption function D_K is not required for decryption in the CFB mode—the encryption function E_K is used for both encryption and decryption. While this reduces the space required to implement the CFB mode, it also makes this mode insecure to be implemented where an adversary might have an access to the encryption function E_K, for example, if the block cipher is a public-key algorithm.

Also, the CFB mode does not require message padding if the message-length is not a multiple of the block size n. If the length of the last block, x_ℓ, is r, then c_ℓ is just the XOR of x_ℓ with the first r bits of $E_K(c_{\ell-1})$ and the rest of the $n-r$ bits of $E_K(c_{\ell-1})$ are discarded.

The main advantage of the CFB mode over the CBC mode is that it can be implemented, with a slight modification (see Figure 7.5), to enable encryption of blocks of length r-bits, $1 \leq r \leq n$. When $r = 1$, this mode encrypts one character at a time and acts as a stream cipher.

Let msb_r (resp. lsb_r) denote the r most (rep. least) significant bits of a block. The plaintext is divided into ℓ r-bit blocks, $x_1, x_2, \ldots, x_\ell$ which output ℓ r-bit ciphertext blocks, $c_1, c_2, \ldots, c_\ell$, each of length r.

■ **Encryption:**
$O_0 \leftarrow IV; c_i \leftarrow \text{msb}_r(E_K(O_{i-1})) \oplus x_i; O_i \leftarrow \text{lsb}_{n-r}(O_{i-1}) \| c_i;$

■ **Decryption:**
$O_0 \leftarrow IV; x_i \leftarrow \text{msb}_r(E_K(O_{i-1})) \oplus c_i; O_i \leftarrow \text{lsb}_{n-r}(O_{i-1}) \| c_i;$

Consider the following example.

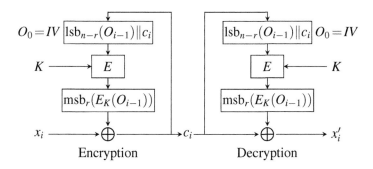

Figure 7.5: CFB Mode with r-bit Feedback

Example 106 *We consider the block cipher, plaintext, ciphertext key space and encryption function as defined in the Example 105. Let the initial vector $IV = 1100$, $r = 3$ and the plaintext is $m = 010100010100110101$. Divide the whole message into fixed 3 bits of plaintext block as $x_1 = 010, x_2 = 100, x_3 = 010, x_4 = 110, x_5 = 101$. Encryption of m using CFB mode is given in the following table*

i	O_{i-1}	I_i	t_i	x_i	c_i
1	1100	0110	011	010	001
2	0001	1000	100	100	000
3	1000	0100	010	010	000
4	0000	0000	000	100	100
5	0100	0010	001	110	111
6	0111	1011	101	101	000

where $I_i = E_\pi(O_{i-1})$, $t_i = msb_r(I_i)$. The resultant ciphertext is $c = 001000000100111000$.

7.2.4 Output FeedBack (OFB)

The output feedback (OFB) mode of operation is a close relative of the CFB mode and shares many properties with the CFB mode. It also makes use of an IV and block chaining to output distinct ciphertexts at each instance. The encryption is almost identical to that of the CBC mode, just that the output of the block cipher encryption E_K is sent back to the E_K instead of the ciphertext block c_i.

■ **Encryption:**
   $O_0 = c_0 \leftarrow IV; O_i \leftarrow E_K(O_{i-1}); c_i \leftarrow O_i \oplus x_i$

■ **Decryption:**
   $O_0 \leftarrow c_0; O_i \leftarrow E_K(O_{i-1}); x_i \leftarrow O_i \oplus c_i$

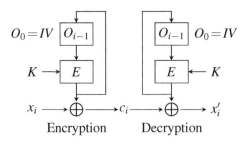

Figure 7.6: Output FeedBack Mode

7.2.4.1 OFB Properties

As in the CFB mode, the OFB mode also has the *chaining dependency* during encryption which causes the loss of the parallelizability and the random access property during encryption. But the decryption does not suffer since the ith plaintext block $x_i = c_i \oplus E_K(O_{i-1})$ can be recovered independently of other blocks as long as the IV, $c_0 = O_0$, and the ith ciphertext block are available. During decryption, the *error propagation* is limited and the decryption process *self-synchronizes* to recover from errors — though an r-bit change/error in a ciphertext block inverts the corresponding bits in the corresponding block of plaintext, the rest of the blocks remain intact as long as the IV, $c_0 = O_0$, was received correctly.

Like the CFB mode, the OFB mode also does not require the block cipher decryption function D_K. The encryption function E_K is used for both encryption and decryption. Also, the OFB mode does not require message padding if the message-length is not a multiple of the block size n. If the length of the last block, x_ℓ, is r, then c_ℓ is just the XOR of x_ℓ with the first r bits of $E_K(O_{\ell-1})$ and rest of the $n - r$ bits of $E_K(O_{\ell-1})$ are discarded.

The OFB mode can also be implemented, with a slight modification (see Figure 7.7), to enable encryption of blocks of length r-bits, $1 \leq r \leq n$. When $r = 1$, this mode encrypts one character at a time and acts as a stream cipher.

Let msb_r (resp. lsb_r) denote the r most (resp. least) significant bits of a block. The plaintext is divided into ℓ r-bit blocks, $x_1, x_2, \ldots, x_\ell$ which output ℓ r-bit ciphertext blocks, $c_1, c_2, \ldots, c_\ell$, each of length r.

■ **Encryption:**
$O_0 \leftarrow IV; O_i \leftarrow E_K(O_{i-1}); c_i \leftarrow \text{msb}_r(O_i) \oplus x_i;$

■ **Decryption:**
$O_0 \leftarrow IV; O_i \leftarrow E_K(O_{i-1}); x_i \leftarrow \text{msb}_r(O_i) \oplus c_i;$

The main advantage of the OFB mode over the CFB mode is that during the OFB mode of operation, the plaintext or the ciphertext is only used for the final XOR. So, the block cipher operations may be performed in advance, allowing the final step to be performed in parallel once the plaintext or ciphertext is available.

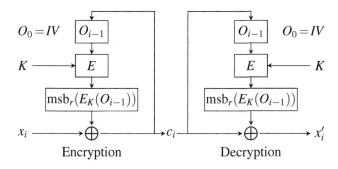

Figure 7.7: OFB Mode with r-bit Feedback

Further, the OFB mode *keystream* can be obtained by using the CBC mode with a constant string of zeroes as input which allows usage of the fast hardware implementations of CBC mode for OFB mode encryption.

While in the CBC and CFB modes, the reuse of the same key and IV combination would leak the similarities while still maintaining some confidentiality, in the case of the OFB mode, this would be catastrophic for security since this will lead to the same keystream being generated. So, the IV in the OFB mode should be used only once, sometimes referred to as *nonce*, even though it can be known to the adversary in advance and need not be unpredictable as required in the CBC and CFB modes. Consider the following example.

Example 107 *We consider the block cipher, plaintext, ciphertexl key space, and encryption function as defined in the example 106 of CFB mode. Let the initial vector $IV = 1100$, $r = 3$, and the plaintext is $m = 010100010100110101$. Divide the whole message into fixed 3 bits of plaintext block as $x_1 = 010, x_2 = 100, x_3 = 010, x_4 = 110, x_5 = 101$. Encryption of m using OFB mode is given in the following table:*

i	O_{i-1}	I_i	t_i	x_i	c_i
1	1100	0110	011	010	001
2	0110	0011	001	100	101
3	0011	1001	100	010	110
4	1001	1100	110	100	010
5	1100	0110	011	110	101
6	0110	0011	001	101	101

where $I_i = E_\pi(O_{i-1})$, $t_i = msb_r(I_i)$. Thus, the resultant ciphertext is $c = 001101110001010101$.

7.2.5 Counter (CTR)

The Counter (CTR) mode of operation is a stream cipher mode similar to the OFB mode with the main difference being in the keystream generation, instead of an iterative encryption of the output of the previous encryption, each keystream element is computed independently of any other keystream element and the encryption at any instance does not depend on the values of any of the plaintexts or ciphertexts. Thus, the CTR mode allows for the random access during encryption and decryption and permits very efficient implementations in software or hardware by exploiting parallelization. As in the OFB mode, the keystream may be pre-computed and the plaintext or the ciphertext blocks, once available, may be XORed with the keystream parallelly implemented once the plaintext or ciphertext is available.

- **Encryption:**
  $$ctr_i \leftarrow ctr + i; c_i \leftarrow x_i \oplus E_K(ctr_i)$$

- **Decryption:**
  $$ctr_i \leftarrow ctr + i; x_i \leftarrow c_i \oplus E_K(ctr_i)$$

- ctr is a binary string of given block size. $ctr + i$ denotes a $|ctr|$-bit string obtained by regarding ctr as a non negative integer, adding i to this number, taking the result modulo $2^{|ctr|}$ and converting this number back into a $|ctr|$ bit string.

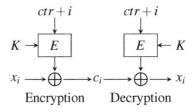

Figure 7.8: Counter Mode

7.3 Padding

If a plaintext is shorter than the block size, additional data is appended to the plaintext in order to make a complete block. Generally, a 1 is appended to the end of the message, starting a new message block if required and then as many 0s are appended as required to form an n-bit or r-bit block. This processing of the plaintext so that the input to the block cipher is of length n is called *padding* and the added string $1000\ldots0$ is called the *pad*.

7.3.1 Ciphertext Stealing

Padding causes ciphertext expansion due to the additional bits appended to the plaintext. Though this is not an issue in the stream cipher modes of operations such as CFB (with 1-bit feedback), OFB, and CTR, in the ECB and CBC modes it can cause upto an n bit expansion. When ciphertext expansion is not desirable, *ciphertext stealing* is a replacement to padding for encrypting messages whose length is not a multiple

Table 7.1: Examples of padding for a 8-bit block cipher

Original Message	Padded Message
1000	10001000
10001000	10001000 10000000
1000100	10001001
10001001	10001001 10000000

of the block size without causing ciphertext expansion, as long as the message is longer than n bits in length since ciphertext stealing requires at least two blocks in the encryption process.

7.3.1.1 Ciphertext Stealing (ECB mode)

Let the last block x_ℓ be of length $|x_\ell| < n$. For $i = 1, \ldots, \ell - 1$, $c_i = E_K(x_i)$ are computed as usual and the second-last ciphertext block $c_{\ell-1}$ is parsed as $c_\ell || P$ where the length of the string c_ℓ is same as that of the last block x_ℓ and the string P is a temporary pad of length $n - |x_\ell|$. P is then appended to x_ℓ to form a full plaintext block $x'_\ell = x_\ell || P$ of length n which is then encrypted and the output ciphertext block is now denoted $c_{\ell-1}$. The temporary pad P is discarded and $c_1, \ldots, c_{\ell-1}, c_\ell$ are sent.

During decryption, for $i = 1, \ldots, \ell - 1$, $x_i = D_K(c_i)$ are computed as usual and the second-last plaintext block $x_{\ell-1}$ is parsed as $x_\ell || P$ where the length of the string x_ℓ is same as that of the last ciphertext block c_ℓ and the string P is a temporary pad of length $n - |c_\ell|$. P is then appended to c_ℓ to form a full ciphertext block $c'_\ell = c_\ell || P$ of length n which is then decrypted and the output plaintext block is the second-last plaintext block $x_{\ell-1}$.

7.3.1.2 Ciphertext Stealing (CBC mode)

Let the last block x_ℓ be of length $|x_\ell| < n$. For $i = 1, \ldots, \ell - 1$, $c_i = E_K(c_{i-1} \oplus x_i)$ are computed as usual where c_0 is the IV. The second-last ciphertext block $c_{\ell-1}$ is parsed as $c_\ell || P$ where the length of the string c_ℓ is same as that of the last block x_ℓ and the string P is a temporary pad of length $n - |x_\ell|$. Let O be a string of $n - |x_\ell|$ 0 bits which is appended to x_ℓ to form a full plaintext block $x'_\ell = x_\ell || O$ of length n. The second-last ciphertext block $c_{\ell-1}$ is then replaced by $E_K(c_\ell || P \oplus x_\ell || O)$. The temporary pad P is discarded and $c_1, \ldots, c_{\ell-1}, c_\ell$ are sent.

During decryption, for $i = 1, \ldots, \ell - 1$, $x_i = c_{i-1} \oplus D_K(c_i)$ are computed as usual and the second-last plaintext block $x_{\ell-1}$ is parsed as $x_\ell || P = c_\ell || O \oplus D_K(c_{\ell-1})$ where the length of the string x_ℓ is same as that of the last ciphertext block c_ℓ and the string P is a temporary pad of length $n - |c_\ell|$ and O be a string of $n - |x_\ell|$ 0 bits. P is then appended to c_ℓ to form a full ciphertext block $c'_\ell = c_\ell || P$ of length n which is then used to output the second-last plaintext block $x_{\ell-1} = c_{\ell-2} \oplus D_K(c_\ell || P)$.

7.4 Design Considerations

Most of today's block ciphers constructions use the basic design principles introduced by Shannon [154] in 1948. He introduced the concept of the *product cipher*, also termed as the *iterated cipher*, which he described as cascades of *diffusion* and *confusion* layers and explained that simple operations, when iterated over "sufficiently" large number of rounds, can provide "sufficient" security.

7.4.1 Diffusion

The *diffusion layer*, also called the *linear layer* or the *P-box*, aims to avoid statistical cryptanalysis by dissipating the *statistical structure* of the plaintext in the long range statistics of the ciphertext, that is, by redistributing the *non-uniformity and redundancy in the statistics* in the distribution of the plaintext into the distribution of much larger structures of the ciphertext in a manner such that bits from all the positions in the original plaintext should contribute to each bit of ciphertext and changing any one bit of the plaintext should change at least half the bits of the ciphertext. It is generally achieved by shuffling or reordering or a linear mixing of bits, by employing permutations of the input, or linear transformations of the Boolean functions or linear codes like MDS (Maximum Distance Separable) codes or similar linear functions.

7.4.2 Confusion

The *confusion layer*, also called the *nonlinear layer* or the *S-box* layer, distributes the key on the plaintext and obscures the statistical relationship between the key and the ciphertext in a manner such that each bit of the key should impact each bit of ciphertext and changing any one bit of the key should change half the bits of the ciphertext. It aims to make the relationship between the plaintext and the ciphertext as complex and involved as possible so that even if one has a large number of plaintext-ciphertext pairs produced with the same key, it is very hard to deduce anything about the key. It is generally achieved by substituting the bits by employing *S-boxes* which are constructed keeping in mind that even complex statistical algorithms cannot find any exploitable structure about the key even from a large number of plaintext-ciphertext pairs.

S-boxes may be constructed in many fashions including using nonlinear Boolean functions, or using mathematical functions such as the power functions $f(x) = x^d$ over a Galois field $GF(2^n)$ for some n, or even generated in a pseudorandom and verifiable way repeatedly until we find an S-box that satisfies all the criteria that we seek, including, better performance of cipher security, satisfaction of the strict avalanche criterion (SAC) (detailed in Subsection 7.4.3), high algebraic and nonlinear degree, efficiency of its implementation, and many others.

7.4.3 Avalanche Effect

One important feature of the iterated ciphers is the *avalanche effect* which refers to the compounding of the small uncertainities added to the plaintext in each round so that the output ciphertext has the property that a small change in either the key or the plaintext causes a significant change in the ciphertext in an unpredictable or pseudorandom manner. That is, even though an adversary can predict the plaintext from the output of one round of the block cipher with a probability $\rho \in [1/2, 1)$ which is very close to 1, after a sufficient number of rounds, ρ should get multiplied to itself so that the probability of predicting any bit of the plaintext correctly should become asymptotically close to $1/2$. Ideally, a block cipher must satisfy the *strict avalanche criterion* (SAC), that is, each of the output bits should change with probability half whenever a single input bit is complemented.

7.4.4 Basic Design Considerations

While designing a block cipher, the required estimated cryptographic security is of paramount importance and the components used in the algorithm must be chosen only if those have been subjected to and withstood expert cryptanalysis over a substantial time period and after an evaluation based on the logarithmic measure of the fastest known computational attack on each of those and in combination with each other. Some of the basic design considerations, other than the computational efficiency, space efficiency, memory requirements during execution, data expansion, throughput, etc, are:

- The block size n impacts the security (larger is more secure against brute force / code book attacks) and the complexity (larger is more costly to implement and causes slower encryption/decryption) of the block cipher.

- The key size k too impacts both the security (larger is desirable) and the complexity (larger is more costly to store and causes slower encryption/decryption) of the block cipher.

- The same is true about the number of rounds of the block cipher which depends directly on the key size k and should be at least $(d \times n)/w$, where d is the number of rounds needed to substitute all the input bits, n is the number of input bits to the block cipher, and w is the number of input bits to the S-box.

- Complexity of the cryptographic mapping in each round, including the key-mixing and the linear and nonlinear layers, affects the implementation costs both in terms of development and fixed resources. So, the tradeoff with the security and the number of rounds must be optimized.

- The *key scheduling algorithm* (KSA) takes as input the user-provided key and outputs a *round key* required for the each round of the block cipher. Though the greater complexity of the KSA will lead to greater difficulty of cryptanalysis, the cost of KSA is fixed and potentially significant, overhead both in terms time and space. When long messages are being encrypted, the cost of the key schedule is amortized but for the encryption of short messages, on-the-fly key generation may be preferred as there is no need to generate and store all the subkeys at the same time which might be desirable in very constrained devices.

7.5 Data Encryption Standard (DES)

The *Data Encryption Standard* (DES) originated in 1973 with IBM's effort to develop banking security systems by a team led by Horst Feistel. The initial version was Lucifer, which admitted a 128-bit key size and block size but was broken. It was modified and submitted in response to NBS' (currently NIST) call [122] for algorithms for protecting sensitive but unclassified data. After further modifications, it was designated the US government FIPS standard in 1977.

Its design was based on a *Feistel Network* which later became the basis for many block ciphers. A Feistel Network is an iterated cipher mapping (L_0, R_0) to (R_r, L_r) through r-round process, $(L_{i-1}, R_{i-1}) \to_{K_i} (L_i, R_i)$ as in Figure 7.9.

$$L_i = R_{i-1}$$

$$R_i = L_{i-1} \oplus f(K_i, R_{i-1})$$

K_i is derived from K

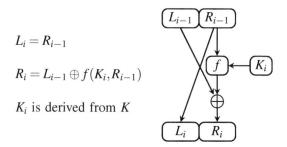

Figure 7.9: The Feistel Network

Example 108 (2 Round Example of a Feistel Network) *The encryption steps are*

$$L_1 = R_0, R_1 = L_0 \oplus f(K_1, R_0) \quad and \quad L_2 = R_1, R_2 = L_1 \oplus f(K_2, R_1) \quad (7.1)$$

and the decryption steps are

$$R_1 = L_2, L_1 = R_2 \oplus f(K_2, R_1) \quad and \quad R_0 = L_1, L_0 = R_1 \oplus f(K_1, R_0). \quad (7.2)$$

The DES takes as input 64-bit plaintext P and 56-bit key K and outputs 64-bit ciphertext C. There are 16 rounds of encryption/decryption for which the round keys are computed using an independent key scheduling algorithm (KSA) (Subsection 7.5.3) which takes as input the key K and outputs sixteen 48-bit subkeys, K_i, $i = 1, \ldots, 16$.

An internal variable state is initialized with the input plaintext P and is updated by the *Initial Permutation IP* described in Table 7.2. The leftmost and rightmost

32-bit subblocks of state are denoted by L_0 and R_0. This state $= IP(P) = L_0 \| R_0$ becomes the input for the first round. For $i = 1, \ldots, 16$, we update the state using

$$L_i = R_{i-1} \quad \text{and} \quad R_i = L_{i-1} \oplus f(K_i, R_{i-1}). \tag{7.3}$$

where f is the *scrambler function*, also called the *mangler function*, described in Figure 7.11 and L_i and R_i are the leftmost and the rightmost 32-bit subblocks of the i-th round output state. L_{16} and R_{16} are swapped and the state is updated by the inverse IP^{-1} of the initial permutation IP described in Table 7.2. Finally, the ciphertext C is defined to be the state. That is, $C = IP^{-1}(R_{16} \| L_{16})$

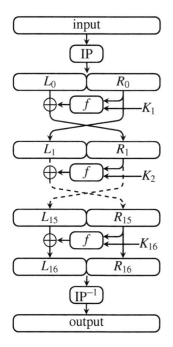

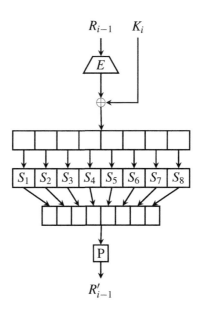

Figure 7.10: The DES Structure   **Figure 7.11:** The Scrambler/Mangler Function

Table 7.2: *IP* and *IP*$^{-1}$

58	50	42	34	26	18	10	2		40	8	48	16	56	24	64	32
60	52	44	36	28	20	12	4		39	7	47	15	55	23	63	31
62	54	46	38	30	22	14	6		38	6	46	14	54	22	62	30
64	56	48	40	32	24	16	8		37	5	45	13	53	21	61	29
57	49	41	33	25	17	9	1		36	4	44	12	52	20	60	28
59	51	43	35	27	19	11	3		35	3	43	11	51	19	59	27
61	53	45	37	29	21	13	5		34	2	42	10	50	18	58	26
63	55	47	39	31	23	15	7		33	1	41	9	49	17	57	25

7.5.1 The Mangler Function f

The *scrambler function*, also called the *mangler function*, f is the main cryptographic mapping of DES and takes as input the 32-bit right half subblock, R_{i-1}, and the 48-bit ith round key, K_i, and outputs a 32-bit block R'_{i-1} defined by

$$R'_{i-1} := f(R_{i-1}, K_i) := P\big(S\big(E(R_{i-1}) \oplus K_i\big)\big) \tag{7.4}$$

where

- E is an *expander* function, described in Table 7.3, which expands R_{i-1} to 48 bits,

- P is a 32-bit permutation, described in Table 7.4, and

- S is the nonlinear component which compresses the 48-bit XOR of E's output with K_i, to output 32-bits through eight S-boxes $S_1, \ldots, S_8$ each of which takes a 6-bit input and has a 4-bit output.

Table 7.3: The Expander Function E

32	1	2	3	4	5
4	5	6	7	8	9
8	9	10	11	12	13
12	13	14	15	16	17
16	17	18	19	20	21
20	21	22	23	24	25
24	25	26	27	28	29
28	29	30	31	32	1

Table 7.4: The Permutation Function P

16	7	20	21
29	12	28	17
1	15	23	26
5	18	31	10
2	8	24	14
32	27	3	9
19	13	30	6
22	11	4	25

7.5.2 The S-boxes

DES has eight S-boxes $S_1, \ldots, S_8$ (described in Table 7.5), each of which takes a 6-bit input and has a 4-bit output. The first and the last bits of the input to box S_i form a 2-bit binary number which indicates one of four rows in the table for S_i. The middle four bits form a 4-bit binary number which indicates one of sixteen columns in the table for S_i. The decimal value in the cell selected by the row and column is then converted to its 4-bit representation to produce the output.

Example 109 *For the first S-box, S_1, let the 6-bit input be* 011011. *Then, from the first and last bits, we get the row index* 01 *and from the middle four bits we get the column index* 1101 = d *so that the output is* 5 = 0101.

S1	0	1	2	3	4	5	6	7	8	9	a	b	c	d	e	f
00	e	4	d	1	2	f	b	8	3	a	6	c	5	9	0	7
01	0	f	7	4	e	2	d	1	a	6	c	b	9	5	3	8
10	4	1	e	8	d	6	2	b	f	c	9	7	3	a	5	0
11	f	c	8	2	4	9	1	7	5	b	3	e	a	0	6	d

Table 7.5: The DES *S*-boxes

S1	0 1 2 3 4 5 6 7 8 9 a b c d e f
00	e 4 d 1 2 f b 8 3 a 6 c 5 9 0 7
01	0 f 7 4 e 2 d 1 a 6 c b 9 5 3 8
10	4 1 e 8 d 6 2 b f c 9 7 3 a 5 0
11	f c 8 2 4 9 1 7 5 b 3 e a 0 6 d

S2	0 1 2 3 4 5 6 7 8 9 a b c d e f
00	f 1 8 e 6 b 3 4 9 7 2 d c 0 5 a
01	3 d 4 7 f 2 8 e c 0 1 a 6 9 b 5
10	0 e 7 b a 4 d 1 5 8 c 6 9 3 2 f
11	d 8 a 1 3 f 4 2 b 6 7 c 0 5 e 9

S3	0 1 2 3 4 5 6 7 8 9 a b c d e f
00	a 0 9 e 6 3 f 5 1 d c 7 b 4 2 8
01	d 7 0 9 3 4 6 a 2 8 5 e c b f 1
10	d 6 4 9 8 f 3 0 b 1 2 c 5 a e 7
11	1 a d 0 6 9 8 7 4 f e 3 b 5 2 c

S4	0 1 2 3 4 5 6 7 8 9 a b c d e f
00	7 d e 3 0 6 9 a 1 2 8 5 b c 4 f
01	d 8 b 5 6 f 0 3 4 7 2 c 1 a e 9
10	a 6 9 0 c b 7 d f 1 3 e 5 2 8 4
11	3 f 0 6 a 1 d 8 9 4 5 b c 7 2 e

S5	0 1 2 3 4 5 6 7 8 9 a b c d e f
00	2 c 4 1 7 a b 6 8 5 3 f d 0 e 9
01	e b 2 c 4 7 d 1 5 0 f a 3 9 8 6
10	4 2 1 b a d 7 8 f 9 c 5 6 3 0 e
11	b 8 c 7 1 e 2 d 6 f 0 9 a 4 5 3

S6	0 1 2 3 4 5 6 7 8 9 a b c d e f
00	c 1 a f 9 2 6 8 0 d 3 4 e 7 5 b
01	a f 4 2 7 c 9 5 6 1 d e 0 b 3 8
10	9 e f 5 2 8 c 3 7 0 4 a 1 d b 6
11	4 3 2 c 9 5 f a b e 1 7 6 0 8 d

S7	0 1 2 3 4 5 6 7 8 9 a b c d e f
00	4 b 2 e f 0 8 d 3 c 9 7 5 a 6 1
01	d 0 b 7 4 9 1 a e 3 5 c 2 f 8 6
10	1 4 b d c 3 7 e a f 6 8 0 5 9 2
11	6 b d 8 1 4 a 7 9 5 0 f e 2 3 c

S8	0 1 2 3 4 5 6 7 8 9 a b c d e f
00	d 2 8 4 6 f b 1 a 9 3 e 5 0 c 7
01	1 f d 8 a 3 7 4 c 5 6 b 0 e 9 2
10	7 b 4 1 9 c e 2 0 6 a d f 3 5 8
11	2 1 e 7 4 a 8 d f c 9 0 3 5 6 b

7.5.3 Key Schedule

The 64-bit key K is input as a 8-byte block of which the bits $8, 16, \ldots, 64$ are to assure that each byte is of odd parity. The DES key scheduling algorithm (KSA) first extracts the 56-bit key K from 64-bit input key data by applying a permutation PC1 : $\{0,1\}^{64} \to \{0,1\}^{28} \times \{0,1\}^{28}$, called Permuted Choice 1 (described in Table 7.6), on it. The result is then split into two 28-bit halves C_0 and D_0 which are then updated as

$$C_i = LS_{r_i}(C_{i-1}) \quad \text{and} \quad D_i = LS_{r_i}(D_{i-1}) \tag{7.5}$$

where LS_{r_i} is the left circular shift by r_i bits and r_i are as follows:

round	1	2	3	4	5	6	7	8	9	10	11	12	13	14	15	16
r_i	1	1	2	2	2	2	2	2	1	2	2	2	2	2	2	1

(7.6)

It then concatenates C_i and D_i and applies a permutation PC2 : $\{0,1\}^{28} \times \{0,1\}^{28} \to \{0,1\}^{48}$, called Permuted Choice 2 (described in Table 7.6), which contracts (drops 4 bits from each half) and then permutes to output the 48-bit round key K_i.

Table 7.6: PC1 and PC2

		14 17 11 24 1 5 3 28
57 49 41 33 25 17 9	63 55 47 39 31 23 15	15 6 21 10 23 19 12 4
1 58 50 42 34 26 18	7 62 54 46 38 30 22	26 8 16 7 27 20 13 2
10 2 59 51 43 35 27	14 6 61 53 45 37 29	41 52 31 37 47 55 30 40
19 11 3 60 52 44 36	21 13 5 28 20 12 4	51 45 33 48 44 49 39 56
		34 53 46 42 50 36 29 32

7.5.4 DES Variants

Many cryptographers believed 56-bit key size was too small for security and proposed variations in the original DES with longer keys to achieve a better brute force security while still enjoying the implementation efficiency provided by DES. Most notable among these variations were DESX and TDES which we describe below.

7.5.5 DESX

DESX was proposed by Ron Rivest in 1984 to increase the security of DES against exhaustive key search. DESX takes as input 64-bit plaintext P and 184-bit key K and outputs 64-bit ciphertext C. K is parsed as $K_1 \| K_2 \| K_3$, where K_1 and K_3 are 64-bit subblocks and K_2 is a 56-bit subblock. The ciphertext C is then given by

$$C := DESX_{(K_1,K_2,K_3)}(P) := K_3 \oplus DES_{K_2}(K_1 \oplus P). \qquad (7.7)$$

By taking $K_1 = K_3 = 0$, DESX results in the original DES to maintain backward compatibility. The step $K_1 \oplus \cdot$ is an example of *pre-whitening* and the step $K_3 \oplus \cdot$ is an example of *post-whitening*. In some implementations, the 184-bit key is derived from a 128-bit key and in some others K_3 is a strong one-way function of K_1 and K_2. Even though the key size is 184, the effective security of DESX is only 120-bits. Further, if 2^x plaintext-ciphertext pairs are available to the attacker, then the attack complexity reduces to $120 - x$.

7.5.6 TDES

It was observed early on that DES was an idempotent cryptosystem. That is, if E_K and D_K denote the encryption and decryption operations of DES with a key K, then for two independent keys K_1 and K_2, the product ciphers defined by $E_{K_2} \circ E_{K_1}$ and $D_{K_2} \circ E_{K_1}$ did not reduce to E_{K_3} for any key K_3. This motivated researchers to consider n-fold products of *DES* to increase the security of DES against exhaustive key search by taking longer keys. But due the 64-bit block size, taking very large n was not productive while for $n = 2$, the meet-in-the-middle attack, proposed by Diffie & Hellman in 1977, reduced the effective security to be of order 2^{56} operations.

In 1999, IBM proposed TDES popularly known as *triple DES* and also denoted by 3DES which was an "encrypt-decrypt-encrypt" product, $E_{K_3} \circ D_{K_2} \circ E_{K_1}$, of DES and somewhat prevented the meet-in-the-middle attack. TDES has block size 64 and admits key sizes 168, 112 (by taking $K_3 = K_1$), and 56 (by taking $K_2 = K_1$) which have effective bit security 112, 80 and 56 respectively. In some implementations, the 168-bit key is derived from a 128-bit key and in some others K_3 is a strong one-way function of K_1 and K_2. Observe that keeping the decryption in the middle step ensures backward compatibility with DES by taking $K_2 = K_1$.

7.6   Advanced Encryption Standard (AES)

On noting the advances in cryptanalysis of the data encryption standard (DES), in 1997, NIST issued a call for proposal [123] for its replacement, the *advanced encryption standard* (AES) [124] which would be a 128-bit block cipher admitting key sizes of 128, 192, and 256 bits. 21 algorithms were submitted of which 15 entries were accepted for the first round. 5 algorithms were selected for the second round from which Joan Daemen and Vincent Rijmen's *Rijndael* [51] was selected as winner and was designated AES after some modifications.

The submissions were required to have a reasonably sound mathematical basis, have a pseudorandom output ciphertext, support variable key sizes and block sizes, implementable on a wide variety of platforms, and must be either unpatented or require no licensing and be available royalty-free world-wide. The submissions were evaluated based on their security as compared with other submissions and security concerns raised by the security community, their computation efficiency on both software and hardware, and their memory requirements, among various other parameters.

The operations in Rijndael were based on finite field $(GF(2^8))$ arithmetic and were implemented using only very simple operations: table lookup, XOR, and shift operations. It is a *substitution-permutation network* (SPN) where layers of S-box substitution and diffusion are iterated over rounds.

Though Rijndael supported many key sizes and block sizes, the AES was standardized to have a block size of 128 bits, key sizes of 128, 192, and 256 bits, and have respective number of rounds (N_r) 10, 12, or 14.

We describe the 128-bit key size version of the AES below. The other versions are similar and details are available in [51].

The 128-bit input (plaintext/key) $b_0 b_1 \ldots b_{127}$ is expressed as 16 bytes $B_0 B_1 \ldots B_{15}$ where $B_i = b_{8i} b_{8i+1} \ldots b_{8i+7}$. These bytes are then represented by 4-byte words (called state) $w_0 w_1 w_2 w_3$ where $w_i = B_{4i} B_{4i+1} B_{4i+2} B_{4i+3}$ which are also represented as the 4×4 array (7.8) with each word w_i forming a column of the array. So, in AES-128 algorithm, a plaintext/ciphertext and input/output of each round looks like array (7.8).

$$
\begin{array}{|c|c|c|c|}
\hline
B_0 & B_4 & B_8 & B_{12} \\
\hline
B_1 & B_5 & B_9 & B_{13} \\
\hline
B_2 & B_6 & B_{10} & B_{14} \\
\hline
B_3 & B_7 & B_{11} & B_{15} \\
\hline
\end{array}
$$

(7.8)

7.6.1 Role of $GF(2^8)$ in AES

As we have discussed a plaintext/ciphertext of AES is a matrix of 4×4 bytes (called state). A byte can be written as a pair of hexadecimal number. For example

Example 110 *A byte 10010001 can be expressed as a pair $\{91\}$ in hexadecimal number. Similarly, a hexadecimal pair $\{BA\}$ corresponds to the byte 10101001.*

In AES, a byte is identified by an element of the finite field $GF(2^8)$ under irreducible polynomial $x^8 + x^4 + x^3 + x + 1$, see Chapter 2 for details. For example, a byte $b = b_7 b_6 b_5 b_4 b_3 b_2 b_1 b_0$ corresponds to the polynomial $\sum_{i=0}^{7} b_i x^i \in GF(2^8)$. So, the multiplication/addition/division of two bytes corresponds to the corresponding operations of two polynomials in $GF(2^8)$. Also, the inverse of a byte corresponds to the inverse of respective polynomial in $GF(2^8)$. Consider the following example.

Example 111 *The byte $b = 00100000$ corresponds to the polynomial x^5 in $GF(2^8)$. As we have seen in Example 24, the inverse of x^5 is equal to $x^5 + x^4 + x^3 + x$ in $GF(2^8)$ which corresponds to the byte 00111010. Thus $(00100000)^{-1} = 00111010$.*

Thus, a state of AES corresponds to 4×4 array of polynomials in $GF(2^8)$.

7.6.2 Basic Steps

An internal variable state is initialized with the input plaintext x on which the following four basic operations are performed in the main encryption algorithm with the *key-scheduling* (KSA) being done independently.

■ SUBBYTES (Figure 7.13)—the confusion/nonlinear step where each byte B_i of state is substituted, using an algebraically defined S-box, which can be computed on the fly or, more commonly, implemented using 16 parallel lookups of the Table 7.8. Mathematically, a byte B of a state is mapped to $B \leftarrow AB^{-1} \oplus C$, where A and C are defined by

$$A = \begin{pmatrix} 1 & 0 & 0 & 0 & 1 & 1 & 1 & 1 \\ 1 & 1 & 0 & 0 & 0 & 1 & 1 & 1 \\ 1 & 1 & 1 & 0 & 0 & 0 & 1 & 1 \\ 1 & 1 & 1 & 1 & 0 & 0 & 0 & 1 \\ 1 & 1 & 1 & 1 & 1 & 0 & 0 & 0 \\ 0 & 1 & 1 & 1 & 1 & 1 & 0 & 0 \\ 0 & 0 & 1 & 1 & 1 & 1 & 1 & 0 \\ 0 & 0 & 0 & 1 & 1 & 1 & 1 & 1 \end{pmatrix} \quad \text{and} \quad C = \begin{pmatrix} 1 \\ 1 \\ 0 \\ 0 \\ 0 \\ 1 \\ 1 \\ 0 \end{pmatrix} \quad (7.9)$$

■ SHIFTROWS (Figure 7.14)—the first part of the diffusion/linear step which reorders the bytes within each row of the array 7.8 by left-rotating the row r_i by i byte positions i.e. left byte shift is done as : first row 0 shift, second row 1 left shift, third row 2 left shift and the fourth row 3 left shift.

■ MIXCOLUMNS (Figure 7.15)—the second part of the diffusion/linear step which reorders the bytes within each column of the array 7.8 by multiplying the column by a 4×4 MDS (Maximum Distance Separable) matrix (Equation 7.10) over $GF(2^8)$. Mathematically, the mixed column transformation can be defined as the product of two polynomials in $GF(2^8)$. Let $w_i = (B_{4i}B_{4i+1}B4i + 2B_{4i+3})$, for $0 \leq i \leq 3$, be the ith column of the state matrix. If we express the column by a polynomial

$$B_{4i} + B_{4i+1}x + B_{4i+2}x^2 + B_{4i+3}x^3 \in GF(2^8)[x]$$

Then, the mixed column transformation is

$$w_i \leftarrow (w_i \star \{03\}x^3 + \{01\}x^2 + \{01\}x + \{02\})mod(x^4 + 1)$$

where $\star$ denotes the convolution product of polynomials with coefficients in $GF(2^8)[x]$.

The mixed column multiplication can also be expressed as $w_i \leftarrow Mw_i$, where

$$M = \begin{pmatrix} 02 & 03 & 01 & 01 \\ 01 & 02 & 03 & 01 \\ 01 & 01 & 02 & 03 \\ 03 & 01 & 01 & 02 \end{pmatrix} \quad \text{and} \quad M^{-1} = \begin{pmatrix} 0e & 0b & 0d & 09 \\ 09 & 0e & 0b & 0d \\ 0d & 09 & 0e & 0b \\ 0b & 0d & 09 & 0e \end{pmatrix}$$

(7.10)

■ ADDROUNDKEY (Figure 7.12)—the key-mixing step where the state is XORed (bytewise) with the roundkey. This is the only step where the encryption/decryption key is used.

Table 7.7: AES Encryption over n Rounds

ADDROUNDKEY	pre-whitening
SUBBYTES	↑
SHIFTROWS	for $n - 1$
MIXCOLUMNS	rounds
ADDROUNDKEY	↓
SUBBYTES	
SHIFTROWS	final round
ADDROUNDKEY	

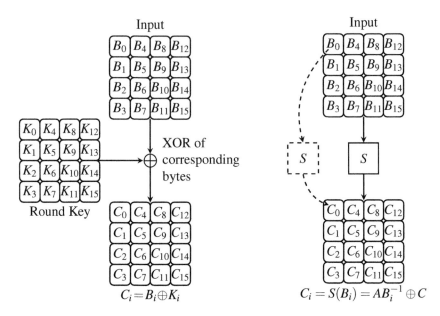

Figure 7.12: ADDROUNDKEY **Figure 7.13:** SUBBYTES

Table 7.8: The AES S-box

	0	1	2	3	4	5	6	7	8	9	a	b	c	d	e	f
0	63	7c	77	7b	f2	6b	6f	c5	30	01	67	2b	fe	d7	ab	76
1	ca	82	c9	7d	fa	59	47	f0	ad	d4	a2	af	9c	a4	72	c0
2	b7	fd	93	26	36	3f	f7	cc	34	a5	e5	f1	71	d8	31	15
3	04	c7	23	c3	18	96	05	9a	07	12	80	e2	eb	27	b2	75
4	09	83	2c	1a	1b	6e	5a	a0	52	3b	d6	b3	29	e3	2f	84
5	53	d1	00	ed	20	fc	b1	5b	6a	cb	be	39	4a	4c	58	cf
6	d0	ef	aa	fb	43	4d	33	85	45	f9	02	7f	50	3c	9f	a8
7	51	a3	40	8f	92	9d	38	f5	bc	b6	da	21	10	ff	f3	d2
8	cd	0c	13	ec	5f	97	44	17	c4	a7	7e	3d	64	5d	19	73
9	60	81	4f	dc	22	2a	90	88	46	ee	b8	14	de	5e	0b	db
a	e0	32	3a	0a	49	06	24	5c	c2	d3	ac	62	91	95	e4	79
b	e7	c8	37	6d	8d	d5	4e	a9	6c	56	f4	ea	65	7a	ae	08
c	ba	78	25	2e	1c	a6	b4	c6	e8	dd	74	1f	4b	bd	8b	8a
d	70	3e	b5	66	48	03	f6	0e	61	35	57	b9	86	c1	1d	9e
e	e1	f8	98	11	69	d9	8e	94	9b	1e	87	e9	ce	55	28	df
f	8c	a1	89	0d	bf	e6	42	68	41	99	2d	0f	b0	54	bb	16

Input

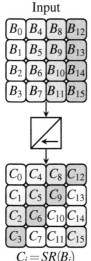

$$C_i = SR(B_i)$$

Figure 7.14: SHIFTROWS

Input

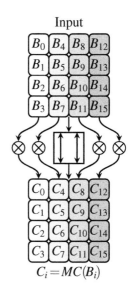

$$C_i = MC(B_i)$$

Figure 7.15: MIXCOLUMNS

7.6.3 Basic Structure

The state is *pre-whitened* with the input key k before the first round. The pre-whitening is sometimes referred to as the 0th round in which the state is updated by performing 0th ADDROUNDKEY operation 0th roundkey, $k_0 = k$. For each of the first $N_r - 1$ rounds, the substitution operation SUBBYTES is performed on state, followed by the permutation SHIFTROWS, the linear transformation MIXCOLUMNS, and the key-mixing operation ADDROUNDKEY. In the final (N_r-th) round, the MIX-COLUMNS operation is skipped and only the operations SUBBYTES, SHIFTROWS, and ADDROUNDKEY are performed. Finally, the ciphertext y is defined to be the state.

7.6.4 AES–Key Schedule

To complete AES-128 encryption/decryption, we generate 10 round keys (i.e. 40 words) from a single 128-bits cipher key using an algorithm called key scheduling algorithm (KSA). The AES key scheduling algorithm(KSA)(also called AES key expansion) is performed using the following three basic operations.

■ ROTWORD$(B_0B_1B_2B_3) = B_1B_2B_3B_0$.

■ SUBWORD$(B_0B_1B_2B_3) = B_0'B_1'B_2'B_3'$ where $B_i' = $ SUBBYTES(B_i).

■ XORWORD$(B_0B_1B_2B_3) = B_0'B_1B_2B_3$ where $B_0' = B_0 \oplus RC[i]$ and $RC[i]$

(round constant) is given by

$$RC[1] = 1$$
$$RC[i] = 2 \times RC[i-1] \text{ for } i > 1$$

In polynomial representation, we can express the round constant as $RC[1] = 1$ and $RC[i] = x \times RC[i-1]$ in $GF(2^8)$.

Thus, 10 round constant are obtained as
$RC[1] = 1 = 00000001, RC[2] = x = 00000010, RC[3] = x^2 = 00000100,$
$RC[4] = x^3 = 00001000, RC[5] = x^4 = 00010000, RC[6] = x^5 = 00100000,$
$RC[7] = x^6 = 01000000, RC[8] = x^7 = 10000000, RC[9] = x^8 = x^4 + x^3 + x +$
$1 = 00011011,$
$RC[10] = x^9 = x^5 + x^4 + x^2 + x = 00110110.$

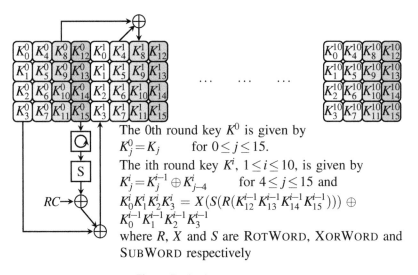

The 0th round key K^0 is given by
$K_j^0 = K_j$ for $0 \le j \le 15$.
The ith round key K^i, $1 \le i \le 10$, is given by
$K_j^i = K_j^{i-1} \oplus K_{j-4}^i$ for $4 \le j \le 15$ and
$K_0^i K_1^i K_2^i K_3^i = X(S(R(K_{12}^{i-1} K_{13}^{i-1} K_{14}^{i-1} K_{15}^{i-1}))) \oplus$
$K_0^{i-1} K_1^{i-1} K_2^{i-1} K_3^{i-1}$
where R, X and S are RotWord, XorWord and SubWord respectively

Figure 7.16: The AES Key Schedule

In terms of words, for a given initial cipher key $K^0 = (w_0, w_1, w_2, w_3)$, the ith round keys $K^i = (w_{4i} w_{4i+1} w_{4i+2} w_{4i+3})$, $i = 1, 2, 3 \ldots 10$ are generated recursively as follows:

1. $w_{4i+l} = w_{4(i-1)+l} \oplus w_{4i+l-1}$, for $l = 1, 2, 3$.

2. $w_{4i} = w_{4(i-1)} \oplus X(S(R(w_{4i-1})))$, where R, S, and X are *RotWord*, *SubWord* and *XorWord* transformation respectively as defined above.

7.6.5 AES–Decryption

Every operation in the AES encryption is invertible and during decryption these inverse operations with some modifications are used in the same order as for encryption with modified round keys.

During decryption, the KSA cannot directly generate round keys in reverse order so either all round keys must be stored or the 'final' state must be pre-computed and then work backwards from that, which means extra time is required from getting key to start of first decryption.

7.7 Exercises

1. Let the block cipher be defined as in definition 7.1. Then, find how many possible E and E_K are there?

2. Let $P_1, P_2, \ldots, P_p \in \mathcal{P}$ and $K \in \mathcal{K}$. Let $\mathcal{K}_i \subseteq \mathcal{K}$, $1 \leq i \leq p$, be defined by

$$\mathcal{K}_i := \{K' \in \mathcal{K} \mid E_{K'}(P_j) = E_K(P_j) \text{ for all } 1 \leq j \leq i\}.$$

   (a) What is the size $|\mathcal{K}_i|$ of the set $\mathcal{K}_i$?
   (b) What is the smallest p for which $|\mathcal{K}_p| = 1$?

3. Decrypt the ciphertext 100111100101 using ECB mode and CBC mode. Use the permutation with block length 3 and the key $\begin{pmatrix} 1 & 2 & 3 \\ 3 & 1 & 2 \end{pmatrix}$. For the CBC mode, use the initialization vector: 100.

4. Decrypt the ciphertext 111100001010101 using OFB mode and CFB mode. Use the permutation with block length 4 and key $\begin{pmatrix} 1 & 2 & 3 & 4 \\ 4 & 1 & 2 & 1 \end{pmatrix}$. Choose the message block length $r = 3$ and the initialization vector as 0100.

5. Encrypt the plaintext 110011001100 using ECB mode, CBC mode, CFB mode and OFB mode. Use the permutation with block length 3 and key $\begin{pmatrix} 1 & 2 & 3 \\ 2 & 3 & 1 \end{pmatrix}$. The initialization vector for CBC, CFB and OFB mode is 001.For OFB mode and CFB mode choose $r = 2$.

6. Show the result of passing 110110 through S-box 2, S-box 3, S-box 6, and S-box 7.

7. In DES prove that the complement of the ciphertext C obtained using the key K and message m is same as the ciphertext obtained by using the complement of the key K with complement of the message m.

8. Why is mixing transformation needed in AES but not in DES?

9. In DES, a weak key is one that, after parity drop operation, consists of all 0s, all 1s or half 0s and half 1s. Explain the disadvantages of weak keys in DES.

10. Let $K = 000\ldots000$ be the DES key consisting all zeros. Show that if $E_K(E_K(P)) = P$.

11. Why is mixing transformation needed in AES but not in DES?

12. Suppose the cipher key of AES consists of all 0s. Find the first and the second round keys of AES-128.

13. Suppose the cipher key of AES consists of all 1s. Find the first, the second and the third round keys of AES-128.

14. Prove that the SubByte transformation of AES is not an affine map.

Chapter 8

Hash Function

CONTENTS

The notion of function has been extensively studied and used in mathematics. The functions are useful in many domains due to their special properties. Implementation and security are prime concerns of modern cryptography. A one-way function, with the compression property is very useful in designing of secure cryptographic primitives. We study this function as *Hash Function* which is commonly used in construction of modification detection code (MDC), message authentication codes (MAC), and digital signature.

8.1 Compression and Hash Functions

Before the discussion of the properties of hash function and their advantages, we first formalize the following simple definitions. We use the notation $\{0,1\}^\star$ for arbitrary length bitstring and $\{0,1\}^n$ for the bitstring of length n.

8.1.1 Compression Function

We say that a function f is a compression function if it maps an input of finite length to an output of fixed shorter length.

Example 112 *Consider a function* $f: \{0,1\}^n \times \{0,1\}^n \to \{0,1\}^n$ *such that* $f(x,y) = x \oplus y$. *This function maps $2n$ bits message to n bits message.*

8.1.2 Computable Map

We say that a map $h: X \to Y$ is a computable map, if it is *easy* to compute the image $y = h(x)$. In the information-theoretic view, *easy* refers to the complexity in polynomial time.

8.1.3 Hash Function

We refer a function h as a hash function , if it maps an input of arbitrary length to an output of fixed length. A hash function is at least compressible and computable.

Example 113 *Consider a function* $h: \{0,1\}^\star \to \{0,1\}^8$, *such that* $h(b_1b_2b_3b_4..........)$ $= B_1 \oplus B_2 \oplus B_3 \oplus ...$, *where B_1 is first byte, B_2 is the second byte from left and so on. If the last B_i is not of eight bits, we add required number of 0's to make it a byte. For example,* $h(10011010100101010101010100001) = 10011010 \oplus 10010101 \oplus 01010101 \oplus 01000010 = 00011000$.

Remark 8.1   Due to compressibility, a hash function is many-to-one function, and, hence, may bring collisions.

8.1.4 Additional Properties

1. **One-way function/First preimage-resistance:** For a function $h : X \to Y$ it is computationally infeasible to find $x = h^{-1}(y)$, for given $y = h(x)$.

2. **Weak collision-resistance/Second preimage-resistance:** For a given pair of (x, y) with $y = h(x)$, under a function $h : X \to Y$, it is computationally infeasible to find a second preimage x', such that $h(x') = y$.

3. **Strong collision-resistance:** For a function $h : X \to Y$, it is computationally infeasible to find two distinct values x, x' (i.e. $x \neq x'$), such that $h(x) = h(x') = y$.

For security purpose of cryptographic primitives, a hash function is suitable in cryptography if with the basic properties of computability and compressibility it also satisfies the above three additional properties, namely preimage-resistance, second preimage-resistance, and collision-resistance. In this view, such hash functions are often termed as *cryptographic hash functions*. For an example, if either of the properties 2 and 3 does not hold then any cryptographic primitive such as MAC or digital signature (we will discuss further in detail) computed over the hashed value of message x will be also valid for the message x' as $h(x) = h(x')$. Hence, integrity and authentication both cannot be guaranteed.

The computability refers existence of a polynomial-time algorithm for computation of hash function, where the compressibility refers that hash function $h : X \to Y$ is defined only for the input $x \in \{0, 1\}^{m(n)}$ and maps $x \in X$ always as $h(x) \in \{0, 1\}^{l(n)}$, where $l(n)$ and $m(n)$ are bitlengths for security parameter n and $m(n) > l(n)$. To prevent the birthday attack we usually choose $l(n)$ of size 160 bits. We have following relation between the three properties.

collision-resistance $\Rightarrow$ second preimage-resistance $\Rightarrow^*$ preimage-resistance

The first implication is straightforward. It also implies that the main challenge in designing of hash functions is designing a collision-resistant hash function[#].

* This is conditional with high probability of existence. For the evidence, let us consider a hash function $h : X \to Y$ which is second preimage-resistant and not preimage-resistant. By computability of hash function, for a given x its hash $h(x)$ can be computed easily. Now since it is not preimage-resistant, hence, a value x' can be obtained such that $h(x) = h(x')$. Intuitively, while difference of the input and output size is very large for a hash function, with high probability it can be achieved that $x \neq x'$ but this contradicts the assumption that h is second preimage resistant, hence the assumption, that h is not preimage-resistant, was wrong.

We say that the above case is conditional as it is possible to define a second-preimage resistant hash function which is not preimage-resistance. The following example from [115] can be evidenced.

$$\alpha(x) = \begin{cases} 0||x & \text{if } x \text{ is } n \text{ bit long} \\ 1||h(x) & \text{otherwise} \end{cases}$$

where $h(x)$ is a collision-resistant hash function.

[#] Remember, finding a collision-resistant hash function requires careful computations and observations to avoid collision finding algorithms by mounting collision with high probability. For example, suppose, the hash function $h : X \to Y$ takes q inputs say $x_1, x_2, .., x_q \in X$ and outputs in Y where $|Y| = n$, then it will reflect a collision with probability 1 if $q > n$. But, there may be collision for a smaller q too. So, the question is how large the value of q is sufficient to meet a collision. The answer is $q = O(\sqrt{|Y|})$. Hence, for a hash function which outputs a 128-bit number, only 2^{64} trials are sufficient to find a collision, though the computation is difficult but not impossible with current computational resources. A popular attack, well studied in cryptography for this purpose, is birthday attack. Thus, for the practical purposes of hash function the output size are usually fixed greater than 128-bit. For example, hash function SHA-1 outputs a 160-bit number. Designing a collision-resistant hash function with arbitrary-length input is another challenge. However, designing a compression function is comparatively easy. For such a purpose, first a compression function (hash function with fixed-length input) can be designed and then it can be converted to an arbitrary-length input hash function using some efficient transformations like Merkle-Damgård Transformation [91].

8.2 Hash Function for Cryptography

This section is the main objective of this chapter. Here, we discuss the need of hash function in cryptography. As mentioned above, application of hash function is majorly realized in achieving integrity of message and authentication of source. Message Authentication Codes (MAC) and digital signature are most useful tools to achieve the message integrity and source authentication. We precisely discuss below the role of hash function in these algorithms. We start with the security goals of cryptography and then identify the role of hash function in offering the required security.

8.2.1 Security Aspects of Communication

Cryptography is often realized to setup secure communication between authorized users. In general it means that the information transferred between the two pre-determined users should be confidential even over the public domain. Encryption is a strong tool to prevent a third party by reading the original message shared between the users. But, unfortunately, this does not encompass full security of the communication. For example, the third party on the communication channel is always able to change the content of the communication, many times he is also able to trap the original content and route the communication through his end with a different content. Keeping in mind these situations, the receiver cannot ensure the originality of the received message by the authentic source, just by performing a decryption. Additionally the receiver also need to perform checks for what is called integrity of the message as well as the authenticity of the source of the message. For an instance, suppose there is communication between Alice and Bob. Alice is CEO of a company and Bob is personal assistant to Alice. In her absence, if Alice sends a message to Bob for taking a confidential and sensitive financial decision for the company, before execution, Bob has to confirm that the message is not tampered and indeed sent by Alice, otherwise it may cause loss to the company. Hence, for more security guarantee, the authentication is appended with the ciphertext.

8.2.2 Modification Detection Code (MDC)

Hash function can be used to maintain the integrity of the message. Suppose, Alice wants to send an important message m to Bob. Alice first computes the hash image $h(m)$ (called message digest or MDC) and send it using a secure channel (e.g. a trusted third party). The message can be sent using any insecure channel. Bob received the pair $(m, h(m))$ and computes the hash image of first component. If hash image of the first component is equal to the second component, then he confirms that the integrity of the message is maintained. If both, the message and its hash image, are sent through insecure channel, then an attacker can modify the message and can create a new MDC for the modified message. MDC is more useful to check the integrity of own message. For example, when Alice writes a program in her computer, she can store hash value of the program. Later, when she again visits the program in her computer, she can compute hash value of the program. If both, the stored and the computed hash values of the program are same, she can confirm that the program has not been tampered meanwhile.

8.2.3 Message Authentication Codes (MAC)

Using MDC, we can only check the integrity of the message. To check integrity and authentication both, a Message Authentication Code (MAC) is used. A MAC is an electronic value, number, or binary string depending upon the specific message and a specific secret key sent along with the message. For a secure communication, users share the message, MAC of the message and a common secret key. Equality of the received MAC and computed MAC (using the received message and the shared secret key) ensures both integrity and authentication. The secret key is shared using some secure channel.

Suppose a communication between Alice (sender) and Bob (receiver) has been established. It has been assumed that during the setup, prior to the communication,

Alice and Bob both have shared the same secret key. The MAC supports the security of communication in the following way:

1. Alice forwards the message appending its MAC with it.

2. Receiving the message and the MAC, Bob computes MAC on the received message using the shared secret key.

3. Bob then checks equality between the received MAC and the computed MAC. On their matching, Bob accepts the message and assures that the message is not tampered and it was actually sent by Alice.

4. If the computed MAC does not match to the received MAC, Bob can ensure that either it was not sent by Alice or the message has been tampered.

8.2.3.1   MAC Algorithm

Formally, a MAC algorithm can be defined by the three-tuple (**KeyGen**, **TagGen**, **Verify**) where:

1. $k \leftarrow$ **KeyGen**(1^n): On input security parameter n, this outputs key k.

2. $t \leftarrow$ **TagGen**(k, m): On input shared key k and the message m, this outputs a tag t.

3. $b \leftarrow$ **Verify**(k, m, t): On input shared key k, message m and tag t, this outputs a bit b, which is 1 if tag is valid, 0 otherwise.

For consistency, it should hold

$$Verify_k(m, \textbf{TagGen}(k, m)) = 1.$$

Remark 8.2   MAC does not bring privacy or confidentiality to the communication. The confidentiality is a different goal of cryptography and can be achieved by encryption. Hence, the message is usually sent as a cleartext in MAC without taking its trap into account during the communication. In practical uses, the encryption is also combined with MAC to attain the required security.

8.2.3.2   Security of MAC

Below we discuss the security setup of MAC.

First, in the security model, the adversary $\mathcal{A}$ has given access to the following oracles:

***KeyGen*:**   On input of security parameter n, this outputs a random key k.

***TagGen*:**   On input of security parameter n, key k and a chosen message m this outputs a tag t on messgae m.

The adversary is allowed to place a polynomial number of queries for the above oracles say q_k for key queries and q_t for the tag queries on its chosen messages. It is also emphasized that the messages for the **TagGen** oracle can be chosen adaptively.

Output: Let, after all the queries, the adversary $\mathcal{A}$ outputs a pair of (message, tag) as (m', t').

Break: Then the adversary is successful in forging the MAC if,

1. $Verify_k(m', \textbf{TagGen}(k, m')) = 1$ where $\textbf{TagGen}(k, m') = t'$.

2. The adversary has never submitted m' in the **TagGen** oracle in the above phase.

Remark 8.3   Here, the first property ensures the validity of the tag (MAC value) of the new unqueried message m', and the second ensures that no original or previously generated valid (m, t) pair has been repeated by the adversary. Essentially, the compulsion of the second property is to save the MAC against the *replay attack*. A replay attack takes place when the previously received (message, tag) pair is repeated with one of the legitimate users. This implies that a replay attack is possible in MAC algorithm. The possible techniques to get rid of this attack include *time stamping* and *sequence numbers*.

However, we concentrate on the security model described above and define advantage of the adversary $\mathcal{A}$ in the above game by the following:

$$Adv_{\mathcal{A},n}^{CMA} = Pr[Verify_k(m', \textbf{TagGen}(k, m')) = 1]$$

where n is security parameter and the advantage is computed against adaptive chosen-message attack (CMA).

More formally:

Theorem 8.1
A MAC algorithm is said to be existentially unforgeable against adaptive chosen-message attack if

$$Adv_{\mathcal{A},n}^{CMA} \leq f(n)$$

where $f(n)$ is negligible function of input security parameter n.

There are four types of MACs: unconditionally secure, hash function-based, stream cipher-based, and block cipher-based. Out of these MACs the hash-based MAC or simply HMAC, which can be viewed as a hash function tailored with the MAC algorithm, is a highly recommended MAC to use.

8.2.3.3 HMAC

In the previous section, we have discussed that a MAC algorithm is existentially unforgeable against adaptive chosen-message attack. But it does not guarantee that the MAC value cannot reveal the content of the message to the adversary. When the message includes some sensitive or private information, then a plain MAC can cause insecurity.

For the purpose of adequate security various cryptographic tools have been appended with MAC. As mentioned in the above note, there are specific cryptographic constructions of MAC. The *H*ash-based *M*essage *A*uthentication *C*ode or HMAC is one of those for such security.

Another advantage of HMAC is that it is a pseudorandom function (PRF) [12, 13]. Hence, when the output is random, guessing an output based on the input or vice versa is infeasible for the adversary in general. Additionally, the HMAC does not require a random *initialization vector* (IV) for security as in MAC, hence, for a specific (bitlength) output, a deterministic input can be used.

So, HMAC is basically a MAC algorithm using the functionality of hash function. Hence, this ensures both integrity of the message and authentication of the source. HMAC is nowadays commonly used for authentication in API calls.

Unlike MAC, in HMAC the user computes message digest using any standard hash function, prior to computing the MAC value of the message. Then, further, the user computes the MAC value of the message digest using an other key k' generated by the original key k.

$$HMAC(k,m) = H(k', H(k', m))$$

where H is the hash function.

8.2.3.4 HMAC Algorithm

Formally, an HMAC algorithm can be defined by the three-tuple (**KeyGen**, **TagGen**, **Verify**) where:

1. $(k, k') \leftarrow \textbf{\textit{KeyGen}}(1^n)$: On input security parameter n, this outputs keys k and k'.

2. $t \leftarrow \textbf{\textit{TagGen}}(k, k', m)$: On input shared keys k, k' and the message m, this outputs a tag $t := h_k((k' \oplus \mathsf{opad}) \| h_k((k \oplus \mathsf{ipad}) \| m))$.

3. $b \leftarrow \textbf{\textit{Verify}}(k, k', m, t)$: On input shared key k, k', message m and tag t, this outputs a bit b, which is 1 if tag is valid, 0 otherwise.

For consistency, it should hold

$$Verify_{k,k'}(m, \textbf{\textit{TagGen}}(k, k', m)) = 1.$$

opad and ipad are two constants with length of the input block size of function h, where opad is formed by repeating the byte "0x36" and ipad by repeating the byte "0x5C" as many times as required.

The input block is with reference to the Merkle-Damgård Transformation [91]. Various hash functions like MD5, SHA-1, SHA-256 and others can be used to construct HMAC. The resulting MAC algorithm is referred to as HMAC-MD5, HMAC-SHA-1, HMAC-SHA-256 accordingly.

8.2.3.5 Limitations of MAC

As in the MAC setup, a common key is required to be shared prior to the communication. This key establishment itself is the bottleneck of MAC. Also, if the symmetric key is established then it ensures that MAC cannot provide the *non-repudiation* property, as both, the sender and the recipient uses the same key, hence any data actually sent by the sender, can be claimed to be generated by the receiver using the common key.

8.2.4 Digital Signature

Above, we had discussed the objective and mechanism of MAC. In general, it provides integrity and authentication. But we also find some limitations of MAC mainly due to the symmetric key used in it.

In Chapter 10, we will discuss that digital signature is an asymmetric-key cryptographic primitive which provides message integrity and source authentication. Hence, the functionality of MAC can be achieved through digital signature also. However, a digital signature does not replace a MAC. The central line of the discussion is importance of hash function in providing integrity and authentication in cryptography by either of the means. In the above section, we investigated the role of hash function in MAC, here, we discuss its role in digital signature.

As the messages may be of the variable lengths and there are certain rules and limitations for input of a message to a digital signature algorithm, in practice, always the digest of message (hashed message) is signed instead of the message itself. Since, the hash function is computable, anyone can compute the digest from the received message and can compare with the signed digest after verifying (decrypting) it. The integrity of the message and authentication of source are verified when the two message digests are same. Any tampering yields difference in the two message digests.

In the above section while discussing the **birthday attack**, we had observed that a hash function can achieve at least the 80-bit security with a 160-bit output, hence, signing a 160-bit (small) document is good idea to save computation and resource overheads.

8.3 Random Oracle Model

The random oracle model (ROM) introduced by Ballare and Rogway [15] is considered as an ideal hash function in cryptography [132]. Hash function is an integral ingredient for security in cryptography. The application of hash function was initially realized to sign long messages with a short signature. Its property of collision resistance helps to achieve non-repudiation. It is observed that hash function appears to be a suitable tool for security when considered to be a random function. This view can be understood by following the idea of random oracle model. In this model, the hash function can be viewed as an oracle which produces a truly random value for each new query i.e. there is no algorithm or formula to reply the query. Note that, for the same query it always responds with the same fixed answer. Proofs in this model guarantee security of the cryptographic scheme, provided that the hash functions are defined without any weakness. Following the model, a hash function is replaced by a random oracle which outputs k bit elements, for k being security parameter of the scheme. This way, level of security achieved by this assumption is 2^k.

8.4 Cryptographic Hash Functions

There is a series of cryptographic hash functions including MD4, MD5, RIPEMD-128, RIPEMD-160, RIPEMD-320, SHA-1, SHA-2 (the set of hash functions SHA-224, SHA-256, SHA-384, SHA-512), SHA-3 (Keccak) etc. Below we briefly discuss three very popular hash functions which have been extensively studied and have also been used for long time. As we know "hash function transforms an arbitrary length bit string to a finite length bit string". It is very difficult to construct such a function directly. So, we use a compression function to construct a hash function, which we call iterated hash function.

8.4.1 Iterated Hash Function

Let $f : \{0,1\}^{n+t} \to \{0,1\}^n$ be a compression function, which transforms $n+t$ bits to n bits (where $t \geq 1$). The following steps are considered to construct a hash function $h : \{0,1\}^* \to \{0,1\}^n$. Let $x \in \{0,1\}^*$ be any bit string of arbitrary length with $|x| \geq n+t+1$.

1. If $|x|$ is not divided by t, then using a padding function append required number of bits so that the length of modified bit string ($y = x||pad(x)$ say) is divided by t. In general, the padding function $pad(x)$ contains binary representation of $|x|$ with required number of zeros. Suppose $y = y_1||y_2||....||y_r$, where $|y_i| = t, \forall i$. The last bit string y_r may contain $pad(x)$ (if required).

2. Choose an initial vector IV (which is a public value) of bitlength m. Let $IV = z_0$. Then compute $z_i = f(z_{i-1}||y_i)$, $0 < i \leq r$.

3. Output $h(x) = z_r$ as the hash image of x.

Remark 8.4 The iterated hash function was initially used in construction of MerkleDamgård Hash function.

8.4.2 Merkle-Damgård Hash Function

The Merkle-Damgård scheme or Merkle-Damgård hash function is a method of designing collision-resistant cryptographic hash functions from collision-resistant compression functions. This design has been used to construct many popular hash algorithms such as MD5, SHA-1, SHA-2, SHA-256 etc.

Let $f : \{0,1\}^{n+t} \to \{0,1\}^n$ (where, $t \geq 2$) be a collision resistant compression function. Using the compression function f, a hash function $h : \{0,1\}^\star \to \{0,1\}^n$ can be constructed as follows. Let $x \in \{0,1\}^\star$,

1. Append a minimum number of zeros from left to x such that the length of the new string (x', say) is divisible by t.

2. Append t zeros in the right hand side of the modified string x', i.e. $x'||0^t$, where 0^t represent t zeros.

3. Determine the binary representation of $|x|$ (length of original string x). Let it be $y = b_1 b_2 b_3 ....b_{k(t-1)-d}$, where $k = \lceil \frac{|x|}{t-1} \rceil$ and $d = k(t-1) - |x|$.

4. Append sufficient zeros (d zeros) to the binary representation of $|x|$ such that the length of modified string is divisible by $t - 1$. Hence, now $y = b_1 b_2 b_3 ....b_{k(t-1)-d}||0^d$. Suppose $y = y_1 y_2 ....y_k$, where y_i is the ith $t - 1$ bits of y.

5. Insert one 1 in front of each y_i, get $z = z_1 z_2 z_3 ... z_k$ where $z_i = 1||y_i$.

6. Finally obtain the padded normalized string as $X = x'||0^t||z$. Suppose $X = X_1 X_2 X_3 ... X_r$, where $X_i \in \{0,1\}^t$, $1 \leq i \leq r$. We call each X_i as a word.

7. To compute hash image of x, set $IV = 0^n$ and then compute the hash image $h(x) = H_r$, where $H_i = f(H_{i-1}||X_i), 1 \leq i \leq r$.

Example 114 *Let $t = 4$ and given $x = 110011001$. Here $|x| = 9$, the binary representation of $|x| = 1001$. Let us do step by step.*

Step. 1, append sufficient number of zeroes to x so that length of binary string of x is divisible by 4, now the modified string is $x' = 000110011001$.

Step. 2, append 4 zeros to x', i.e. $x'||0000$.

Step. 3, binary representation of $|x| = 1001$ (remember $|x| = 9$), append $d = 2$ zeros so that $|x|$ is divided by 3, hence $y = 100100$, here $y_1 = 100$, $y_2 = 100$.

Step. 4, insert a '1' in front of each y_i, to get $z = 11001100$. The padded normalized string is $X = x'||0^4||z = 000110011001||0000||11001100$, clearly $X_1 = 0001, X_2 = 1001, X_3 = 1001, X_4 = 0000, X_5 = 1100, X_6 = 1100$.

Remark 8.5 For the security perspective, a typical choice is $t = 512$ and $n = 128$. It is better to choose $n \geq 160$ to prevent from birthday attack.

Theorem 8.2
The Markle-Damgård scheme is collision resistant if the compression function f is collision resistant.

Proof 8.2. Let us assume that the hash function is not collision resistant, we then show that the original compression function is also not collision resistant.

Assume that $h(x) = h(x')$ for some $x \neq x'$. Let the padded normalized string of x and x' be X and X' respectively, where $X = X_1 X_2 X_3 ... X_r$ and $X' = X'_1 X'_2 X'_3 ... X'_{r'}$. Let the iterated compression values be $H_0, H_1, H_2 ... H_r$ and $H'_1, H'_2, ... H'_{r'}$. Without loss of generality assume that $r \leq r'$. According to our assumption, we have $H_r = H'_{r'}$. Now we have two cases.

Case-I- Let $H_{r-i} = H'_{r'-i}$ for some i such that $0 \leq i \leq r$ but $H_{r-i-1} \neq H'_{r'-i-1}$. Then clearly, $H_{r-i-1}||X_{r-i} \neq H'_{r'-i-1}||X'_{r'-i}$ and $f(H_{r-i-1}||X_{r-i}) = H_{r-i} = H'_{r'-i} = f(H'_{r'-i-1}||X'_{r'-i})$. Thus we get a collision of f.

Case-II- Assume that $H_{r-i} = H'_{r'-i}$, where $0 \leq i \leq r$. To find a collision of f, we need to find $X_{r-i} = X'_{r'-i}$ for some $i \in \{0, 1, 2 ... r\}$ so that $f(H_{r-i-1}||X_{r-i}) = H_{r-i} = H'_{r'-i} = f(H'_{r'-i-1}||X'_{r'-i})$. Thus, we get a collision of f. Further, the question is how to find such $X_{r-i} = X'_{r'-i}$ for some $i \in \{0, r-1\}$? For this we have three cases.

1. If the number of words required to represent the length of x (say $z_1 z_2 z_3 ... z_k$) is less than the number of words required to represents the length of x' (say, $z'_1 z'_2 z'_3 ... z'_{k'}$) i.e. $k < k'$. Then, the word X_{r-k} is the zero string but the words $X'_{r'-k}$ start with '1' bit. Thus we have $X_{r-k} \neq X'_{r'-k}$.

2. If $k = k'$, where k and k' are the number of words required to represent the length of x and x' but $|x| \neq |x'|$, which implies that the binary representation of the length of x and x' will be different in at least one bit. Hence, $Z_i \neq Z'_i$ for at least one i. And hence, $X_{r-i} \neq X'_{r'-i}$ for at least one i.

3. If $|x| = |x'|$ but $x \neq x'$. If $x \neq x'$, then clearly the modified representation of x and x' will have different words with same position. In this case, $k = k'$ and $X_{r-i} = X'_{r'-i}$ for $i = 0, 1, 2, 3, ...k$ but $X_{r-k-i} \neq X'_{r'-k-i}$ for some i,

8.4.3 *MD5*

MD5 stands for Message Digest algorithm 5th version. It was designed by Ronald Rivest in 1991 as an improvement over a previous similar algorithm named MD4.

The MD5 algorithm converts a message of arbitrary length into a message digest of 128 bit length. MD5 uses the Merkle-Damgård construction for the purpose. The MD5 algorithm takes input in 512 bit blocks. In each block, it divides the 512 bit into 16 words $M_0, M_1, ..., M_{15}$, each composed of 32 bit. Before hashing, a padding is always performed and then the algorithm starts hashing of individual words round by round, in a sequence.

8.4.3.1 *Working Principle of MD5*

Practically, the MD5 algorithm works as follows:

1. The message is separated in individual letters. Each letter is represented by its unique ASCII code.

2. The corresponding ASCII numbers are written in binary form.

3. Further, all the binary representations are written together and counted. They are further divided into blocks of 512 bit.

4. A padding is always done with the block such that it is congruent to 448 modulo 512 (i.e. exactly 64 fewer than 512). To do so, first the number 1 is appended to the end and followed by zeros such that the total number of bits are 448 modulo 512. For example, if the block contains 350 bit, then at 351th place 1 is posed and 97 zeros are added after that so that total bits are 448 (which is also 448 modulo 512).

5. Lastly, when the block is left with 448 bit, the length of the original message is represented in 64 bit and added to the last, thus, finally the block is equipped with total 512 bit.

6. Now the block of 512 bit is divided into 16 words of 32 bit each.

7. Further, the block of 512 bit (16 words of 32 bit each) is written as 64 words of 32 bit each by applying some well-defined functions.

8. These 64 words are processed separately by their input in 4 different 'states' (also referred as 'round' sometimes) in a loop. A state processes a message input of 32 bit with 128 bit (4×32) variables A, B, C, and D. Similarly, these 64 words are processed in 64 rounds in the group of 16 rounds at each state.

9. To start the first round, in first state, 4 variables A, B, C, and D (of 32 bit each) are initialized and updated after each completed round.

10. Each state is processed with a non-linear function. There are 4 possible non-linear functions namely F, G, H and I to be used in 4 different states.

- $F(B,C,D) = (B \wedge C) \vee (\neg B \wedge D)$
- $G(B,C,D) = (B \wedge D) \vee (C \wedge \neg D)$
- $H(B,C,D) = B \oplus C \oplus D$
- $I(B,C,D) = C \oplus (B \vee \neg D)$

11. Further, each 32 bit word (message segment) is processed as shown in the Figure 8.1, where M_i is 32-bit block of the message input and K_i is a 32-bit constant, which is different for each operation. $<<< s$ denotes a left bit rotation (shifting) by s places (s varies for each operation) and $\boxplus$ denotes addition modulo 2^{32}.

12. After the completion of the last round, the output values of A, B, C, and D are updated by adding the corresponding initial values of A, B, C, and D and the final output is written together as ABCD which is a 128 bit number.

8.4.3.2 Attacks on MD5

Though, the MD5 has become an extremely popular hash function in cryptography and possibly still remains in use with certain conditions, there have been wide analysis and attacks on MD5. Theoretically, it was not found resistant to the collision. In fact, it was observed to be attacked by brute-force method. We have already discussed in the above section that a ***birthday attack*** can be possible with output of 128 bit value. Few of the notable vulnerabilities include collision attack by Wang et al. [170], preimage attack (faster than the brute-force) by Sasaki et al. [146] and others. Since all the previous attacks were applied to multi-block construction of MD5, in 2010, Xie et al. [175] proposed the construction of the single-block (512-bit) MD5 to get rid of the previous attacks. But soon, Marc Stevens [161] mounted a collision attack on the single-block hash function.

Usually, a hash function is used for security in digital signature, but in 2012, Microsoft realised fake digital signature due to collision in MD5 using Flame malware.

8.4.4 SHA-1

Secure Hash Algorithm-1 (SHA-1) is designed by the National Security Agency USA, motivated by the previously designed hash algorithms MD4 and MD5. The working procedure is almost same as those of MD4 and MD5 with some additional functionalities and rounds. SHA-1 has been considered as a NIST standard. As discussed in previous section, one of the weak points of MD5 was 128 bit output (in the view of birthday attack for collision). The SHA-1 outputs 160 bit message digest and, hence, provides safe extension for brute-force attack.

SHA-1 has been widely used in many security applications and protocols such as TLS, SSL, PGP, SSH, IPsec and others.

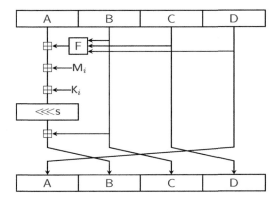

Figure 8.1: MD5 Algorithm

Though there are many similarities among MD4, MD5, and SHA-1, they are a bit different from each other in the following views:

1. In SHA-1, each state consists of 20 operations/rounds instead of 16 like MD4 and MD5, hence, the total rounds in SHA-1 are 80 unlike 64 in MD5 and 48 in MD4.

2. SHA-1 uses 5 words (32 bit each), unlike 4 words of MD4 and MD5.

8.4.4.1 Working Principle of SHA-1

The elementary steps of SHA-1 algorithm are same as those of MD5. Once the padding is done and the length of the message is added as the last 64 bit of the final 512 bit block of message, the further steps are processed as follows:

1. Each 512 bit block is divided into 16 words of 32 bit each.

2. The 16 words are expanded in 80 words by a small function $w_t = <<< 1(w_{t-3} \oplus w_{t-8} \oplus w_{t-14} \oplus w_{t-16})$, for $t = 16, 17, 18...79$. They can be listed from W_0 to W_{79}.

3. There are total 80 rounds, where each ith round takes input of corresponding W_i word.

4. At first, 5 variables are initialized as A, B, C, D, and E. Further, they are updated after each completed round.

5. There is a round function F_t for each round. The 80 words are processed in set of 4 different states. One state consists of 20 rounds taking input of 20 corresponding words. Different round functions F_t ($t = 1,2,3,4$) are used for each state, where

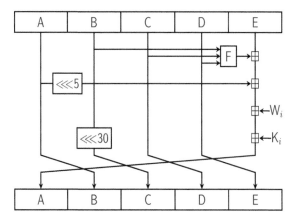

Figure 8.2: SHA-1 Algorithm

- $F_1(B,C,D) = (B \wedge C) \vee (\neg B \wedge D)$
- $F_2(B,C,D) = (B \oplus C) \oplus D$
- $F_3(B,C,D) = ((B \wedge C) \vee (B \wedge D)) \vee (C \wedge D)$
- $F_4(B,C,D) = (B \oplus C) \oplus D$

6. The working process of one round of one state is shown in the Figure 8.2. Where W_i is 32-bit block of the word (message) input, and K_i is a 32-bit constant, which is different for each state. The constant $K_t = 5A827999$ for $t = 1$ to 20, $K_t = 6ED9EBA1$ for $t = 21$ to 40, $K_t = 8F1BBCDC$ for $t = 41$ to 60 and $K_t = CA62C1D6$ for $t = 61$ to 80 respectively. $<<< 5$ denotes a left bit rotation (shifting) by 5 places, similarly $<<< 30$ denotes a left bit rotation (shifting) by 30 places and $\boxplus$ denotes addition modulo 2^{32}.

7. After the completion of round 80, the output values of A, B, C, D, and E are updated by adding the corresponding initial values of A, B, C, D, and E and the final output is written together as ABCDE which is a 160 bit number.

8.4.4.2 Example

Find the digest of the word CRYPTO using SHA-1.

First separately write each letter i.e.   C   R   Y   P   T   O

Now, write ASCII code for the letters i.e.   67   82   89   80   84   79

The binary representation of the ASCII code is

01000011   01010010   01011001   01010000   01010100   01001111.

Hence, binary representation of the message is: 010000110101001001011001010100 00010101010001001111. They are total 48 bits, hence, to make it multiple of 448 bit it is required to append it with additional 400 bit. For that, first we pose '1' at the 49th place and then add 399 zeros after that. Hence, the appended message bit is

0100001101010010010110010101000001010100010011111000000000000000
0000000000000000000000000000000000000000000000000000000000000000
0000000000000000000000000000000000000000000000000000000000000000
0000000000000000000000000000000000000000000000000000000000000000
0000000000000000000000000000000000000000000000000000000000000000
0000000000000000000000000000000000000000000000000000000000000000
0000000000000000000000000000000000000000000000000000000000000000
0000000.

Finally, append the 64 bit for the message length at last. Message length is 48 which is 110000 in binary form. Hence, the 512 bit of message is:

0100001101010010010110010101000001010100010011111000000000000000000000
0000000000000000000000000000000000000000000000000000000000000000000000
0000000000000000000000000000000000000000000000000000000000000000000000
0000000000000000000000000000000000000000000000000000000000000000000000
0000000000000000000000000000000000000000000000000000000000000000000000
0000000000000000000000000000000000000000000000000000000000000000000000
0000000000000000000000000000000000000000000000000000000000000000000000
000000000000000110000.

Since there are total 512 bits finally, hence, there will be only one chunk of 512 bits. Now, divide it to 16 words of 32 bits each as follows:

W_0 : 0100001101010010010110010101010000

W_1 : 0101010001001111100000000000000000

W_2 : 0000000000000000000000000000000000

W_3 : 0000000000000000000000000000000000

W_4 : 0000000000000000000000000000000000

W_5 : 0000000000000000000000000000000000

W_6 : 0000000000000000000000000000000000

W_7 : 0000000000000000000000000000000000

W_8 : 0000000000000000000000000000000000

W_9 : 0000000000000000000000000000000000

W_{10} : 0000000000000000000000000000000000

W_{11} : 0000000000000000000000000000000000

W_{12} : 0000000000000000000000000000000000

W_{13} : 0000000000000000000000000000000000

W_{14} : 0000000000000000000000000000000000

W_{15} : 0000000000000000000000000000110000

Further, expand it to 80 words of 32 bits each. For this, select some particular words and apply some functions among them and output a 32 bit word. First, select W_{13}, W_8, W_2 and W_0 and output $W_{16} = 10000110101001001011001010100000$ by applying $((W_{13} \oplus W_8) \oplus W_2) \oplus W_0$ and left shift by 1 place. Similarly, obtain all 80 words and list them as:

W_0 : 010000110101001001011001010000

W_1 : 010101000100111110000000000000000

W_2 : 000000000000000000000000000000000

W_3 : 000000000000000000000000000000000

W_4 : 000000000000000000000000000000000

W_5 : 000000000000000000000000000000000

W_6 : 000000000000000000000000000000000

W_7 : 000000000000000000000000000000000

W_8 : 000000000000000000000000000000000

W_9 : 000000000000000000000000000000000

W_{10} : 000000000000000000000000000000000

W_{11} : 000000000000000000000000000000000

W_{12} : 000000000000000000000000000000000

W_{13} : 000000000000000000000000000000000

W_{14} : 000000000000000000000000000000000

W_{15} : 000000000000000000000000000110000

W_{16} : 100001101010010010110010101000000

W_{17} : 101010001001111100000000000000000

W_{18} : 000000000000000000000000001100000

W_{19} : 000011010100100101100101010000001

W_{20} : 010100010011111000000000000000001

.

.

.

W_{75} : 101010111100111010000000011101010

W_{76} : 101001100101100011000101000011000

W_{77} : 000100011010011011001100100001001

W_{78} : 001000011001100011111001101001111

W_{79} : 101011101010011000011010010010101010

Now, finally there have been exactly 80 words of 32 bits each, hence, now the SHA-1 algorithm can be applied for 80 rounds. Consider the initialized A, B, C, D, and E as:

$$A = 01100111010000101001000110000001$$
$$B = 11101111110011011010101110001001$$
$$C = 10011000101110101101110011111110$$
$$D = 00010000001100100101010001110110$$
$$E = 11000011110100101110000111110000$$

and follow the round operation as shown in Figure 8.2 and consider F_t as described above for all the four stages. Lastly, after the completion of 80th round, we obtain

$$A = 01001000110111111100010101110000$$
$$B = 01110100001011111100111101010011$$
$$C = 00110011000100000011011101001100$$
$$D = 11101101111111110011010010100000$$
$$E = 00000110010011010010110110100100$$

which gives the values of truncated A, B, C, D, and E, by adding the corresponding values of initial A, B, C, D, and E (which were considered in the very first round) as follows.

$$A = 10110000001001001110100001110001$$
$$B = 01100011111111101011110101011100$$
$$C = 11001011110010110001010001001010$$
$$D = 11111110001100011000100100010110$$
$$E = 11001010001000000000111110010100$$

which are A = b024e871, B = 63fd7adc, C = cbcb144a, D = fe318916 and E = ca200f94 in hexadecimal form.

Hence, the message digest for the word CRYPTO is **b024e87163fd7adccbcb144afe318916ca200f94**.

8.4.4.3   Attacks on SHA-1

SHA-1 serves a better option of hashing in comparison to MD5 as finding a collision is much harder in SHA-1 by the brute-force approach. But practically, the brute-force attack is not the only way of analysis. For example, in 2005 Wang et al. [169] have mounted collision attack in full SHA-1 performing fewer than 2^{69} operations. Recently, many high profiled corporates like Google, Apple and others have announced that they would not use SHA-1 in their browsers for SSL certificates due to a recent practical collision observation in SHA-1. Since last couple of years, organizations have recommended to replace SHA-1 with SHA-2 or SHA-3.

8.5 Exercises

1. How to construct a collision resistant hash function from a collision resistant compression function.

2. Let p be a prime number, $q = (p-1)/2$ also a prime number, a a primitive root mod p, and b be randomly chosen in $\{1, 2, 3, \ldots p - 1\}$. Consider the following map $h : \{0, 1, 2, \ldots q - 1\}^2 \to \{1, 2, \ldots p - 1\}$ such that $h(x_1, x_2) = a^{x_1} b^{x_2} \bmod p$. Prove that h is collision resistant compression function if DLP in $Z_p^{\star}$ is intractable.

3. Construct Merkle-Damgård padding scheme for the case $t = 1$.

4. The Merkle-Damgård scheme is collision resistant if the compression function is collision resistant. Discuss with examples.

5. Using encryption function, explain how to construct a hash function.

6. What is difference between MDC and MAC.

7. The output of a hash function is of n bits. Assuming hash function as a random oracle describe a Las-Vegas type algorithm to find preimage and second preimage. Discuss the success probability of each of the algorithm.

8. Find the label of difficulty in preimage attack, second preimage attack, and collision attack for an ideal hash function.

Chapter 9

Public Key Cryptosystem

CONTENTS

The security of classical ciphers (such as Caesar cipher, Vigenère cipher etc.) depends on the fact that the entire encryption process is secret. If the encryption process is available to the attacker, the attacker may use the frequency analysis to break the cipher. The modern symmetric cipher (such as DES and AES) follow the Kerckhoff's principle (please refer to Chapter 1). According to the principle, the encryption algorithm is known to everyone and the security of the cryptosystem is based on keeping the secret key confidential (as in the symmetric key cryptosystem, encryption key = decryption key). Thus, to communicate using a symmetric key cryptosystem, the user first needs to share the secret key. Now, one would wonder that how should the secret key be shared so that it does not get trapped by any authorized party. A good idea is to share the secret key via some secure (considered to be secure) channel or through some private meeting. Firstly, no channel is secure, even a trusted party, secondly, a the private meeting may not be feasible always (like geographical constraints). An efficient solution proposed for sharing the secret key is *public key cryptography* (PKC). It provides a secure way of sharing secret information (e.g. secret key) even over an insecure channel (public network). The concept of PKC was introduced by Diffie and Hellman in 1976 in their seminal paper "New Directions in Cryptography" [54]. This paper introduced the concept of PKC by providing an ingenious method to exchange a secret key. In an extended view, this work also opened a platform for digital signature. The security of this Diffie-Hellman key exchange protocol is based on the intractability of the discrete logarithm problem (DLP). Although, in [54] the authors proposed the concept of public key cryptosystem but they had no realization of a public-key encryption scheme. In 1978, Rivest, Shamir, and Adleman discovered the first practical public-key encryption and signature scheme, now referred to as RSA cryptosystem [142]. The security of RSA cryptosystem is based on the intractability of factoring large integers. After the invention of RSA cryptosystem, the number theorists and cryptographers pay more attention towards the factoring problems and, hence, proposed many algorithms to solve them. After RSA, many public key cryptosystems were proposed by distinguish cryptographers using different number theoretic problems, for example, ElGamal cryptosystem [58] using DLP, Rabin cryptosystem [136] using quadratic residue problem etc.

9.1   Introduction

The idea behind the public key cryptosystem is that it might be possible to find a cryptosystem where it is computationally infeasible to determine decryption key d for a given encryption key e. If so, then the encryption key e could be made publicly available by publishing in a directory (hence, the term, public key cryptosystem). The advantage of public key cryptosystem is that anyone can encrypt the message with e, while only a legitimate receiver having the secret key d can decrypt it. Alternatively, the private key d can be used for digital signature and anyone can use the corresponding public key to verify its authenticity. The trapdoor one-way function plays a central role in the construction of any public key cryptosystem. We can define the trapdoor one-way function as follows:

Definition 9.1   One-Way Function: A function $f : X \to Y$ is said to be a one-way function if it is easy to compute $f(a)$ for each $a \in X$ but it is very hard, computationally infeasible, to compute a for a given $f(a)$. Here, the term "very hard" or "computationally infeasible" means "determining $a \in X$ such that $b = f(a)$, where b is given, may take billions of years by using all the possible computer power we have". It is difficult to find a one-way function.

Definition 9.2   Trapdoor One-Way Function: A trapdoor one-way function is a one-way function with some secret information (trapdoor). With the help of this trapdoor, anyone can find $a \in X$, such that $b = f(a)$ for any given $b \in Y$ (or can show that such an a does not exist).

As a simple example from everyday life, consider the Delhi telephone directory book. It is easy to find the phone number of any specific person. However, given a specified phone number, it is difficult to find the person with this number by just using the directory. In this sense, the function $f : Persons \rightarrow Phonenumber$ can be visualized as a one-way function.

Following are some mathematical examples of one-way/trapdoor one-way functions.

Example 115 *When n is large, n = pq is a one-way function. Given p and q, it is always easy to calculate n; given n, it is very difficult to compute p and q. This is the factorization problem.*

Example 116 *Given a large prime p and a primitive root g modulo p, it is easy to compute $A = g^a$ (mod p) for a given positive integer $a < p$, but is difficult to compute a for a given g^a (mod p). This is the discrete logarithm problem (DLP).*

Example 117 *Given n = pq, product of two large primes p and q, and an integer e such that $\gcd(e, \phi(n)) = 1$, it is easy to compute $y = x^e$ (mod n). But, it is infeasible to compute x for a given $y = x^e$ (mod n). This is called RSA problem. If someone knows the factor of n or an integer d such that $ed = 1$ (mod $\phi(n)$) then they can compute x for a given $y = x^e$ (mod n). It is a trapdoor one-way function. The factors of n or the integer d is the trapdoor in this RSA problem.*

9.1.1 Symmetric Key Cryptosystem vs. Public Key Cryptosystem

In comparison to symmetric key cryptosystem, public key cryptosystem is slower and requires a longer key (for example, RSA is around one thousand times slower than DES) but it simplifies key distribution problem. Therefore, public key algorithm is commonly used for the *hybrid schemes*: public key system for key distribution and symmetric key system for encryption of the message.

We can differentiate both cryptosystems as follows:

1. *Secret Key*: In symmetric key cryptosystem, $e = d$, but in asymmetric key cryptosystem $e \neq d$, where e is public key and d is secret key.

2. *Number of keys*: If n users use symmetric key cryptosystem, they need $\frac{n(n-1)}{2}$ secret keys. But an asymmetric key cryptosystem requires just n secret keys for n users.

3. *Cryptography Goal*: Symmetric key cryptosystem provides only the first three goals of cryptography, namely, privacy, authenticity and integrity, but using asymmetric key cryptosystem, one can also achieve the fourth goal of cryptography, namely, non-repudiation.

4. *Encryption/Decryption technique*: The encryption/decryption technique of symmetric key cryptosystem is based on substitution and permutation of symbols/binary string, whereas asymmetric key cryptosystem technique follows mathematical functions on numbers.

5. *Security*: Security of asymmetric key cryptosystem is based on some hard mathematical problem, whereas the security of a symmetric key cryptosystem is because of the confidentiality of the secret key.

Let us first describe the well-known Diffie-Hellman key exchange protocol.

9.2 Diffie-Hellman Key Exchange Protocol

Suppose Alice and Bob want to share a secret key for their symmetric cipher but the only means of communication is public or an insecure channel. Every piece of information that they exchange is observed by their adversary Eve. How is it possible for Alice and Bob to share a key without making it available to Eve? Using the concept of discrete logarithm problem (DLP), Diffie and Hellman proposed a brilliant idea to solve this problem. It follows as:

■ The first step is for Alice and Bob to agree on a large prime p (and hence the finite field F_p) and a generator g of F_p which generates a group of large order. In general g is chosen as primitive root modulo p. Alice and Bob make the values of p and g publicly available; for example, they might post the values on their web sites, so Eve too knows. For security reasons (discussed later) it is better if they choose g such that its order in $F_p^\star$ is a large prime.

■ The next step is for Alice to pick a secret integer $a \in_u [2, p-1)$ that she does not reveal to anyone, while at the same time Bob picks an integer $b \in_u [2, p-1)$ that he keeps secret. Bob and Alice use their secret integers to compute $g_b = g^b \pmod{p}$ and $g_a = g^a \pmod{p}$ respectively.

■ Public key for Bob is g_b (including p and g) and the secret key for Bob is b.

Public key for Alice is g_a (including p and g) and the secret key for Alice is a.

■ Next, Alice sends g_a to Bob and Bob sends g_b to Alice using any public channel. Note that, Eve gets to see the values of g_a and g_b since they are sent over the insecure communication channel.

■ Finally, Bob and Alice both compute the common secret key $k = g^{ab} \pmod{p}$ as follows:

Alice computes $k = g_b^a \pmod{p}$ using her secret key a and Bob computes $k = g_a^b \pmod{p}$ using his secret key b. See Algorithm 17 for DH protocol.

In the DH-protocol, users need to generate a primitive root g of p. This can be computed using Algorithm 18.

Let us do the following example.

Algorithm 17 Diffie-Hellman Key Exchange Protocol

COMMON INPUT: (p, g), p is a large prime, g is a generator element in $F_p^\star$.
OUTPUT: An element in $F_p^\star$ shared between Alice and Bob.

1. Alice chooses $a \in_U [1, p-2]$; computes $g^a = g_a \pmod{p}$; sends g_a to Bob.

2. Bob picks $b \in_U [1, p-2]$; computes $g^b = g_b \pmod{p}$; sends g_b to Alice.

3. Alice computes $k = g_b^a \pmod{p}$;

4. Bob computes $k = g_a^b \pmod{p}$.

Algorithm 18 Primitive Root Modulo Prime

INPUT: p, a prime; $q_1, q_2, \ldots, q_k$ all factors of $p - 1$.
OUTPUT: g, a random primitive root modulo p.
Primitive _ root $(p, q_1, q_2, \ldots, q_k)$

1. Pick $g \in_U [2, p-1)$;

2. For $(i = 1, i++, k)$ do
   If $g^{(p-1)/q_i} = 1 \pmod{p}$
   Return (Primitive _ root $(p, q_1, q_2, \ldots, q_k)$).

3. Return (g).

Example 118 *Let $p = 43$. Applying Algorithm 18, Alice and Bob compute the primitive root $g = 3$, p and g are common public key for both Alice and Bob.*

■ *Alice chooses $a = 8$, computes $g_a = 3^8 \pmod{43} = 25 \pmod{43}$. Alice's public key is 25 and her secret key is 8.*

■ *Bob chooses $b = 37$, computes $g_b = 3^{37} \pmod{43} = 20 \pmod{43}$. Bob's public key is 20 and his secret key is 37.*

■ *Alice sends $g_a = 25$ to Bob and Bob sends $g_b = 20$ to Alice.*

■ *Alice computes the shared secret key as $(20)^8 \pmod{43} = 9$ and simultaneously Bob computes the shared secret key as $k = 25^{37} \pmod{43} = 9$. Thus, the common shared secret key is 9.*

Remark 9.1 Choose p, such that $p - 1$ has a sufficiently large prime factor $q(q > 2^{160})$ (necessary to prevent from Pohlig-Hellman Algorithm).

Remark 9.2 g need not be a generator of $F_p^\star$ itself; but it must be a generator of a large order ($= q$) subgroup of $F_p^\star$. In this case, Alice and Bob should check $g \neq 1$ and $g^q = 1 \pmod p$. For this purpose; q should be part of the common input to the protocol.

Remark 9.3 Alice (Bob) should check $g^b \neq 1 (g^a \neq 1)$. Then for their respective exponent chosen from $(1, q)$, these checking steps will guarantee that the shared key g^{ab} would be an element of a large order subgroup.

Remark 9.4 Alice (Bob) should erase her exponent a (his exponent b) and the shared secret key g^{ab} upon termination of the protocol to achieve forward secrecy property.

The Diffie-Hellman key exchange protocol does not support the authenticity of the key agreed. So, an active adversary (Malice) in the middle of the communication between Alice and Bob can manipulate the protocol to succeed in an attack called man-in-the-middle-attack.

9.2.1 The Man-in-the-Middle Attack

The steps of this attack are:

1. Alice picks $a \in_u [2, p-2]$, computes $g_a = g^a \pmod p$; she sends g_a to Malice, assuming that she is sending it to Bob (Malice impersonates himself as Bob);

2. Malice (impersonating as Alice) computes $g_m = g^m \pmod p$ for some $m \in [2, p-2]$;

3. Malice (impersonating as Alice) sends g_m to Bob;

4. Bob picks $b \in_u [2, p-2]$, computes $g_b \in g^b \pmod p$; he sends g_b to Malice, assuming that he is sending it to Alice (Malice impersonates himself as Alice);

5. Malice (impersonating as Bob) sends g_m to Alice;

6. Alice computes $k_1 = g_m{}^a \pmod p$; (k_1 is shared between Alice and Malice, since Malice can compute $k_1 = g_a{}^m \pmod p$));

7. Bob computes $k_2 = g_m{}^b \pmod p$. (k_2 is shared between Bob and Malice, since Malice can compute $k_2 = g_b{}^m \pmod p$)).

9.2.2 CDH Assumption & DL Assumption

Security of any public key cryptosystem is based on some assumptions. The security of Diffie-Hellman key exchange protocol is based on the computational Diffie-Hellman (CDH) assumption. Let us first define the CDH problem and discrete logarithm problem (DLP).

Definition 9.3   Computational Diffie-Hellman Problem: The secrecy of the agreed shared key in the Diffie-Hellman key exchange protocol depends on the computation of g^{ab} (mod p) given g^a and g^b. This problem is called Computational Diffie-Hellman Problem (CDH Problem).

If the CDH problem is easy, then g^{ab} (mod p) can be computed from the values p, g, g^a, g^b, which are transmitted as part of the protocol messages.

Definition 9.4   Discrete Logarithm Problem (DLP): Given a finite field F_p, a generator $g \in F_p^\star$ and an element $h \in F_p^\star$ to find a unique integer $a < p$ such that $h = g^a$ (mod p).

We denote the integer $a = log_g h$.

The CDH-problem and DL-problem are two assumed intractable problems. The difficulty of these problems depends on the size of the problem (by size, we mean the size of F_p) as well as the choice of the parameter g and the private keys a and b.

Definition 9.5   Computational Diffie-Hellmann (CDH) Assumption: If there does not exist any probabilistic polynomial time algorithm to solve the given instance of the Diffie-Helmann problem with non-negligible success probability, we say that the given instance of Diffie-Hellman problem satisfies the Computational Diffie-Hellmann Assumption.

Definition 9.6   Discrete Logarithm Assumption: If there does not exist any probabilistic polynomial time algorithm to solve the given instance of DLP with non-negligible success probability, we say that the given instance of DLP satisfies the Computational Discrete Logarithm Assumption.

These two assumptions state that in finite fields of sufficiently large size, there does not exist any efficient algorithm to solve CDH problem or the DL problem for almost all instances. Holding DL assumption implies the existence of one-way function $g^k : Z_p \rightarrow F_p^\star$. It is still not known whether the function $g^k : Z_p \rightarrow F_p^\star$ is a trapdoor function or not. If the modulus p is composite then we can take the factor of p as a trapdoor.

9.2.2.1   Relation between CDH Assumption & DL Assumption

It is clear that the availability of $a = log_g g_a$ or $b = log_g g_b$ will permit the calculation of $g^{ab} = g_a^b = g_b^a$, which implies that an efficient algorithm that solves the DLP will lead to an efficient algorithm to solve CDH problem. So, we can say that CDH problem is weaker than DLP or CDH assumption is a stronger assumption then DL assumption.

The converse of this is an open question: can DL assumption be true if CDH assumption is false? Maurer & Wolf [113] provided with a strong heuristic argument regarding the relation between these two problems; they suggest that these two problems are equivalent.

9.3 RSA Cryptosystem

In 1978, Rivest, Shamir, and Adleman introduced the first practical and the most popular public-key cryptosystem, the so-called RSA cryptosystem [142]. This system is based on the difficulty of factoring large composite numbers. It can briefly be described as follows.

Key Generation: Bob carefully selects two large primes p and q, computes $n = pq$ and $\phi(n) = (p-1)(q-1)$, chooses an encryption key $e < \phi(n)$ such that $\gcd(e, \phi(n)) = 1$, and computes the decryption key d such that $ed = 1$ (mod $\phi(n)$). The public key of Bob is the pair (n, e) and his private/secret key is d. The plaintext space is Z_n.

Encryption: To encrypt the message $m \in Z_n^{\star}$, Alice computes $c = m^e$ (mod n) using the public key of Bob and sends the ciphertext c to Bob.

Decryption: After receiving the ciphertext c, Bob gets the plaintext by computing $m = c^d$ (mod n).

Correctness of the RSA Cryptosystem:

Since $ed = 1$ (mod $\phi(n)$), we have $ed = k \times \phi(n) + 1$ for some integer k, and hence, $c^d = m^{ed}$ (mod n) $= m^{1+k\phi(n)}$ (mod n) $= m$ (by Euler's theorem). The equation $ed = 1$ (mod $\phi(n)$) is called RSA equation.

Remark 9.5 In general, we take the plaintext and ciphertext space of RSA cryptosystem as Z_n, but the plaintext m must be an element of $Z_n^{\star}$, otherwise $\gcd(m, n)$ will give a factor of RSA modulus n.

Let us do an example.

Example 119 *Bob chooses two primes $p = 23$ and $q = 31$, sets $n = pq = 713$ and computes $\phi(n) = (p-1)(q-1) = 22 \times 30 = 660$. Bob chooses an encryption key $e = 19$ which satisfies $\gcd(e, \phi(n)) = 1$. Next, he computes the private key $d = 139$ such that $ed = 1$ (mod $\phi(n)$) using Extended Euclidean Algorithm. The public key of Bob $= (713, 19)$ and the secret key is 139. The plaintext space and the ciphertext space are $\mathcal{P} = \mathcal{C} = \{0, 1, 2, 3 \ldots, 712\}$ and the key space is $\mathcal{K} = (e : e < 660 \,\& \gcd(e, 660) = 1)$.*
Suppose, Alice wants to send a message $m = 15$ to Bob.
Encryption: Using Bob's public key 19, Alice computes the ciphertext

$$525 = 15^{19} \pmod{713}.$$

and sends the ciphertext 525 to Bob.

Decryption: *After receiving the ciphertext 525, using the private key 139, Bob computes the plaintext*

$$15 = 525^{139} \pmod{713}.$$

Example 120 *Bob chooses two prime numbers $p = 257$ and $q = 337$ and computes the RSA modulus $n = 257 \times 337 = 86609$ and $\phi(n) = 256 \times 336 = 86016$. Bob chooses a public key $e = 17$ and computes the secret key $d = 65777$, such that $ed = 1 \pmod{\phi(n)}$. The public key for Bob is $(86609, 17)$ and the private key is 65777.*

Suppose, Alice wants to send message "HE" to Bob. The numerical representation of "HE" is "0705" (using 00-26 encoding technique). Thus, the message is $m = 0705$ (As per the modulus size, we can encrypt maximum two characters at a time.)

Encryption: *Alice computes the ciphertext $c = 705^{17} \pmod{86609} = 46293$. She sends the ciphertext 46293 to Bob.*

Decryption: *Using the private key 65777, Bob computes the message $m = 46293^{65777} \pmod{86609} = 705$. Thus, the plaintext is $0705 = $ "HE".*

9.3.1 RSA as a Block Cipher

The original message M may be of any arbitrary length. To encrypt the original message using RSA, we first divide the original message into fixed sized blocks $M = m_1 m_2 m_3 \ldots m_k$ such that $m_i < n$, for all $i = 1, 2, 3 \ldots, k$. Then, we use any block cipher technique to encrypt the message block-wise.

Let the RSA modulus be n. Suppose, the plaintext and ciphertext contain N alphanumeric-symbol characters. We use the alphabet $\Sigma = \mathbb{Z}_N = \{0, 1, 2, 3, \ldots N - 1\}$ to represent the plaintext/ciphertext characters. For example, if the plaintext and ciphertext only contains $\{a, b, c, d, \ldots, z, 0, 1, 2, 3, \ldots, 9, ?, space, \&\}$, then $N = 39$ and hence $\Sigma = \mathbb{Z}_{39}$.

In this case, we can define RSA cryptosystem as a block cipher with block length k, where $k = \lfloor log_N n \rfloor$.

Any plaintext word $m_1 m_2 m_3 \ldots m_k \in \Sigma^k$ of block length k corresponds to an integer

$$m = \Sigma_{i=1}^{k} m_i N^{k-i}$$

According to the choice of k, we have

$$0 \le m = \Sigma_{i=1}^{k} m_i N^{k-i} \le (N-1)\Sigma_{i=1}^{k} N^{k-i} = N^k - 1 < n$$

Thus, we can identify each block $m_1 m_2 \ldots m_k$ in Σ^k into an integer $m < n$. The block is encrypted by computing $c = m^e \pmod{n}$. Then, the ciphertext integer c is written in base N as follows:

$$c = \Sigma_{i=0}^{k} c_i N^{k-i}.$$

Thus, $c_0 c_1 c_2 \ldots c_k$ is the ciphertext block corresponding to the plaintext block $m_1 m_2 \ldots m_k$. The N-adic expansion of c may have at most $k + 1 c_i's$. Thus, RSA maps blocks of length k to a block of length $k+1$. This is not a block cipher in the sense of the definition of block cipher as the plaintext and the ciphertext blocks should be of equal length. This block version can be used to slightly modify the version of ECB mode and CBC mode. In any PKC (for example, RSA), it is impossible to use CFB mode or OFB mode due to the same encryption and decryption key in this mode of operation.

Let us take an example to understand the concept.

Example 121 *Consider* $p = 23, q = 31, n = 713, e = 19$ *and* $d = 139$ *as given in the Example 119. Let* $\Sigma = \{0, a, b, c, d\}$ *with numerical identification*

$$
\left(
\begin{array}{c|c|c|c|c}
0 & a & b & c & d \\
\hline
0 & 1 & 2 & 3 & 4
\end{array}
\right).
$$

We obtain $k = \lfloor \log_5 713 \rfloor = 4$, *which is the length of the plaintext blocks. Let a plaintext be* $abc0$ *that corresponds to the block* 1230, *which corresponds to the integer,*

$$ m = 1 \times 5^3 + 2 \times 5^2 + 3 \times 5^1 + 0 \times 5^0 = 190, $$

this is encrypted as

$$ c = 190^{19} \pmod{713} = 225. $$

We express $c = 225$ *in base 5, and obtain*

$$ c = 0 \times 5^4 + 1 \times 5^3 + 4 \times 5^2 + 0 \times 5^1 + 0 \times 5^0. $$

Thus, the ciphertext block is $0ad00$. *If* $c > 5^4$, *then definitely the first left alphabet of the ciphertext will be non-zero. Thus, for a given plaintext "abc0" block of length 4, we get a ciphertext "0abd0" block of length 5.*

9.3.2 RSA Assumption and RSA Problem

The security of RSA cryptosystem is based on RSA assumption and RSA problem as defined below.

Definition 9.7   RSA Problem: The RSA problem is to compute m, for a given $c = m^e \pmod{n}$, where $n = pq$, product of two large primes and e is a positive integer s.t. $e < \phi(n) \& \gcd(e, \phi(n)) = 1$.

Definition 9.8   RSA Assumption: If there exists no probabilistic polynomial time (PPT) algorithm $\mathcal{A}$ to solve RSA-problem with some non-negligible success probability ε, then we that RSA-Assumption holds. That is, no PPT algorithm $\mathcal{A}$ exists such that $Prob[m \leftarrow \mathcal{A}(n, e, m^e \pmod{n})] < \varepsilon$ for some non-negligible quantity $\varepsilon > 0$.

Remark 9.6   Holding of the RSA assumption implies the existence of trapdoor one-way function $m^e : Z_n^\star \to Z_n^\star$ (prime factorization of n is trapdoor). The probability space in this assumption includes the instance space, the plaintext message space, and the space of the random operations of randomized algorithm for solving the RSA problem.

Remark 9.7   Various researches have been conducted to find the relation between both the problems. The relation analyzed in the papers *Breaking RSA may not be equivalent to factoring* [33], *Breaking RSA May Be As Difficult As Factoring* [40] and *Breaking RSA May Be Easier Than Factoring* [32] are interesting to understand the relation between the above two problems and assumptions.

9.3.3   Cryptanalytic Attacks on RSA

In RSA, numerous attacks were proposed by distinguished cryptographers. We will discuss a few of them in this section.

9.3.3.1   Factoring Attack

The first attack on RSA cryptosystem is factoring attack. If the RSA modulus is factored, there is no security at all. Cryptanalysts can get the private key easily. To find the factors of a given composite number is called *Factoring Problem*. If no polynomial time algorithm exists to solve the given instance of factoring problem with non-negligible success probability, then we say that the given instance holds the Factoring Assumption. Many algorithms exist in the literature to solve factoring problem. Some algorithms of factoring has been discussed in Chapter 3. Thus, by using the algorithm for factoring problem we can also solve the RSA problem. But the converse is an open problem till now. Hence, the security of RSA is not equivalent to factoring.

9.3.3.2   Secrete Key and Factoring

Although RSA problem is not equivalent to factoring, but computing secret key d of RSA is computationally equivalent to factorization of RSA modulus n. See the following theorem.

Theorem 9.1
Let $< e, n >$ be an RSA public key. Given the private key d, one can efficiently factor the modulus $n = pq$. Conversely, given the factorization of $n = pq$, one can efficiently recover d.

Proof. First part: The first part is straightforward. Assume that we can factor the RSA modulus $n = pq$, we can compute $\phi(n) = (p-1)(q-1)$ and hence, can compute the secret key d such that $ed = 1 \pmod{n}$ using Extended Euclidean Algorithm.

Converse part: Assume that we have the secret key d. Let $ed - 1 = 2^s k$, where k is an odd number. We can factor the RSA modulus n using Algorithm 19. The algorithm is based on certain facts concerning square roots of 1 modulo n, where $n = pq$ is the product of two distinct odd primes. If $x^2 = 1 \pmod n$, then there are four square roots of 1 modulo n. Two roots are trivial ± 1, other two roots are non-trivial; they are also negative of each other modulo n. If x is a non-trivial square root of 1 modulo n, then $x^2 = 1^2 \pmod n$ but $x \neq \pm 1 \pmod n$. Then, we can find the factors of n by computing $\gcd(x + 1, n)$ and $\gcd(x - 1, n)$ (see Example 122). We will try to find this x using Algorithm 19. This algorithm is based on following two lemmas.

Lemma 9.1

For all integers a those are prime to n, the order of the element a^k in $Z_n^\star$ is in $\{2^i : 0 \leq i \leq s\}$.

Proof. Since $ed - 1 = 0 \pmod{\phi(n)}$, we have $a^{ed-1} = 1 \pmod n \Rightarrow a^{2^s k} = 1 \pmod n \Rightarrow (a^k)^{2^s} = 1 \pmod n. \Rightarrow Ord(a^k)|2^s$. Thus, order of a^k is some integer power of 2.

Lemma 9.2

Let a be an integer that is prime to n. If the orders of the element a^k in $\mathbb{Z}_p^\star$ under modulo p and orders of the element a^k in $\mathbb{Z}_q^\star$ under modulo q are different, then $1 < \gcd(a^{2^t k} - 1, n) < n$ for some $t \in \{0, 1, 2, \ldots, s\}$.

Proof. By Lemma 9.1, the order of the element a^k in $\mathbb{Z}_p^\star$ and a^k in $\mathbb{Z}_q^\star$ are some power of 2, that is, in $\{2^i : i = 0, 1, 2, \ldots, s\}$. Without loss of generality, assume that $ord(a^k) \pmod p > ord(a^k) \pmod q$. Let the order of $a^k \pmod q = 2^t, 0 \leq t \leq s$, then $a^{2^t k} = 1 \pmod q$ but $a^{2^t k} \neq 1 \pmod p$ which implies that $a^{2^t k} \neq 1 \pmod n$. Therefore, $\gcd(a^{2^t k} - 1, n) = q$. In this way, we can find the factor of n. Thus, to find factor of n we proceed as follows:

1. Choose $a \in_u \{1, 2, \ldots, n - 1\}$.

2. Compute $x = \gcd(a, n)$, if $x \neq 1$, x is a factor n.

3. If $x = 1$, then compute $x = \gcd(a^{2^t k} - 1 \pmod n, n)$ for $t = 1, 2, \ldots, s$. If for any a, we find $a^{2^t k} \pmod n \neq \pm 1$ but $a^{2^{(t+1)} k} \pmod n = 1$, then we get a non-trivial square root of 1 modulo n. For this value of t, we get $x > 1$.

4. If $x > 1$, then either $x = p$ or $x = q$. Hence, the factorization of n is found and the algorithm terminates. Otherwise, the algorithm fails for the chosen base a.

5. If the algorithm fails, we repeat the above process for a different chosen a in the set $\{1, 2, \ldots, n - 1\}$.

The success probability of the algorithm is at least $1/2$ (see Theorem 9.2). Therefore, the probability of success after r iteration is at least $1 - \frac{1}{2^r}$.

Algorithm 19 RSA Factor (n, e, d)

INPUT: e, d, n, (Comments-we are assuming that $ed = 1 \pmod{\phi(n)}$)

OUTPUT: p and q, such that $n = pq$

Write $ed - 1 = 2^s k$, k odd

Choose a at random such that $1 < a < n - 1$

$x \leftarrow \gcd(a, n)$

if $1 < x < n$

then return (x)

(**Comment**: x is a factor of n)

$v \leftarrow a^k \pmod{n}$

if $v = 1 \pmod{n}$

then return ("failure")

while $v \neq 1 \pmod{n}$

$$\mathbf{do} \begin{cases} \quad v_0 \leftarrow v \\ \quad v \leftarrow v^2 \pmod{n} \end{cases}$$

If $v_0 = -1 \pmod{n}$

Then return ("failure")

$$\mathbf{else} \begin{cases} x \leftarrow \gcd(v_0 + 1, n) \\ Return(x) \end{cases}$$

(**Comment**: x is a factor of n)

Remark 9.8   If secret key d is revealed (accidently or otherwise), then it is not sufficient to choose a new encryption exponent; one must also choose a new modulus.

Theorem 9.2

The number of integers a in the set $\{1, 2, 3, ...n\}$ which are relatively prime to n and for which a^k has a different order modulo p and q is at least $\dfrac{(p-1)(q-1)}{2}$.

Example 122 *Consider $n = 403 = 13 \times 31$. Four square roots of 1 modulo 403 are -1, 92, 311, and 402. The square root 92 is obtained by solving the system $x = 1 \pmod{13}$ and $x = -1 \pmod{31}$ using CRT. The other square root is $301 = 403 - 91$ (solving $x = -1 \pmod{13}$ and $x = 1 \pmod{31}$). If x is a non-trivial square root of 1 $\pmod{n}$, then $x^2 = 1^2 \pmod{n}$ but $x \neq 1 \pmod{n}$. Two factors of n are $\gcd(x+1, n)$ and $\gcd(x-1, n)$, that is, $\gcd(93, 403) = 31$ and $\gcd(212, 403) = 13$.*

Example 123 *To explain Algorithm 19, we consider $n = 89855713$, $e = 34986517$, $d = 82330933$, and the random value $a = 5$.*

1. *$ed - 1 = 2^3 \times 360059073378795$;*

2. *$a^t \pmod{n} = 85877701$;*

3. *$85877701^2 = 1 \pmod{n}$;*

4. $x = \gcd(85877702, n) = 9103.$

5. *Other factor is* $x = \gcd(85877700, n) = 9871.$

6. *Thus,* $n = 9103 \times 9871.$

9.3.3.3 Computing $\phi(n)$

The secret key d is the inverse of encryption key e under modulo $\phi(n)$, so computing $\phi(n)$, we can easily compute the secrete key from the public key (e,n). Here, we will show that computing $\phi(n)$ is no easier than factoring RSA modulus n. For example, suppose, n and $\phi(n)$ is known, then we have two equations:

$$n = pq$$

and

$$\phi(n) = (p-1)(q-1)$$

Putting $q = \frac{n}{p}$ in the second equation, we have

$$p^2 - (n - \phi(n) + 1)p + n = 0 \tag{9.1}$$

a quadratic equation. Solving this equation, we get two roots p and q, factors of n. Hence, if the cryptanalyst has the value of $\phi(n)$, he can compute the factors of n and break the system.

Consider the following example for illustration.

Example 124 *Let* $n = 84773093$, *suppose we have the value* $\phi(n) = 84754668$. *Now, putting the value of n and* $\phi(n)$ *in equation 9.1, we get*

$$p^2 - 18426p + 84773093 = 0.$$

On solving this equation, we get two roots as $p = 9539$ *and* $q = 8887$, *two factors of n.*

9.3.4 Common Modulus Attack

Suppose, we are hiring a trusted third party (TTP) to generate keys. To avoid generating a different modulus $n = pq$ for different user, TTP may wish to fix n once and the same n can be used by all the users. Finally, TTP could provide the unique encryption and decryption key (e_i, d_i) to ith user from which user i forms a public key (e_i, n) and a private key d_i. At a first glance this may seems to work: a ciphertext $C = M^{e_a} \pmod{n}$ intended for Alice cannot be decrypted by Bob since Bob does not possess d_a. However, this is incorrect and the resulting system is insecure. As we know that, getting private key d is computationally equivalent to factor n. So, using his private key d_b and public key e_b, Bob can factor the modulus n and can compute the private key d_a of Alice.

If a plaintext is encrypted twice with the RSA system using two public RSA keys (e,n) and (f,n) such that $\gcd(e,f) = 1$, $c_e = m^e \pmod{n}$ and $c_f = m^f \pmod{n}$. The plaintext m can be recovered from two ciphertexts. Using Extended Euclidean Algorithm, compute x and y such that $ex + fy = 1$ and then compute

$$c_e^x \times c_f^y = m^{(ex+fy)} = m \pmod{n}.$$

Remark 9.9   A common modulus must not be among a group of users.

9.3.3.5   Low Encryption Exponent Attack

To reduce encryption or signature verification time, it is desirable to select a small encryption exponent e, such as $e = 3$. Suppose, a group of entities have the same encryption exponent e, however, each entity in the group must have its own distinct modulus. If an entity A wishes to send the same message M to three entities whose public modulus are N_1, N_2, N_3 and encryption key is $e = 3$, then A would send $C_i = M^3 \pmod{N_i}$ for $i = 1,2,3$. Since these moduli are most likely pair-wise relatively prime, an eavesdropper observing C_1, C_2, C_3 can use CRT to find a solution x such that $x < N_1 N_2 N_3$; to three Congruences

$$X = C_1 \pmod{N_1} \tag{9.2}$$
$$X = C_2 \pmod{N_2} \tag{9.3}$$
$$X = C_3 \pmod{N_3} \tag{9.4}$$

Since $M^3 < N_1 N_2 N_3$, it must be the case that $x = M^3$. Hence, by computing the integer cube root of x, the eavesdropper can recover the plaintext M.

Remark 9.10   To prevent against such an attack is to pad the pseudorandom bit string. The pseudorandom bit string should be independently generated for each encryption. This process is sometimes referred to as *salting* the message.

Remark 9.11   Small encryption exponents are also a problem for small message M, because if $M < N^{1/e}$, then M can be recovered from the ciphertext $C = M^e \pmod{n}$ simply by computing the integer eth root of C. Salting plaintext messages also circumvents this problem.

Remark 9.12   The value $e = 2^{16} + 1 = 65537$ (a prime) is recommended for the encryption exponent. When the value $e = 2^{16} + 1$ is used, signature verification or encryption requires 17 multiplications.

9.3.3.6   Small Decryption Exponent Attack

As was the case in encryption exponent, one may wish to use small decryption exponent to improve the efficiency of decryption time. However, an attack due to M

Wiener [171] shows that the choice of a small d can lead to a total break of the system. Wiener has shown that if n is the modulus and d is the private exponent, with $d < 1/3n^{1/4}$, then given the public key (e,n) an attacker can efficiently recover d. Boneh and Durfee [27] have improved the bound $d < N^{.292}$.

Remark 9.13 For a typical 1024-bit RSA modulus, the private exponent d should be at least 300 bits in length.

Remark 9.14 If the public exponent e is chosen to be 65537 (the most commonly used value) and we calculate d as $ed = 1 \pmod{n}$, then we are guaranteed to have d nearly as long as n.

9.3.3.7 Meet-in-the-Middle Attack

Due to productive properties of RSA cryptosystem, the Meet-in-the-middle attack was proposed by Boneh, Joux and Nguyen [30] under certain conditions. The condition is: let $c = m^e \pmod{n}$ such that Eve (attacker) knows $m < 2^l$. With non-negligible probability, m is a composite number satisfying $m = m_1 m_2$, with $m_1, m_2 < 2^{l/2}$.

Eve performs the following steps

1. Builds a sorted database $\{1^e, 2^e, 3^e, \dots, 2^{\frac{l}{2}e}\} \pmod{n}$.

2. Searches through the sorted database and tries to find $\dfrac{c}{i^e} \pmod{n}$ (for $i = 1, 2, \dots, 2^{l/2}$) and $\dfrac{c}{i^e} = j^e \pmod{n}$.

3. Eve uncovers $m = ij \pmod{n}$.

Remark 9.15 Never use RSA to encrypt a short key or a password which is less then 2^{64}.

9.3.3.8 Forward Search Attack

If the message space is small or predictable, an adversary can decrypt C by simply an encryption of all the possible plaintext messages until C is obtained.

Remark 9.16 Salting the message or using redundancy function (for example, $m \to m\|m$) can prevent this attack.

9.3.3.9 Cyclic Attack

Let $C = M^e \pmod{n}$, be a ciphertext. Since encryption is a permutation on the message space $1, 2, \ldots, N - 1$, there exists an integer k such that

$$C^{e^k} = C \pmod{n}$$
$$\Rightarrow C^{e^{k-1}} = M \pmod{n}.$$

Thus, an adversary computes $C^e \pmod{n}, C^{e^2} \pmod{n}, \ldots, C^{e^k} \pmod{n}$ till he gets C. Then, the penultimate step gives the plaintext.

9.3.3.10 Partial Key Exposure Attack

An attack on the RSA cryptosystem due to Boneh, Durfee and Frankel [25] shows the importance of protection of the entire private key d. If modulus N is k bit long, given $k/4$ least significant bits of d, an attacker can reconstruct all of d in time linear to $(e \log e)$, where e is the public exponent. This means if e is small, a quarter of bits of d can lead to the recovery of the whole private key d.

9.3.3.11 Timing Attack

Timing attack takes advantage of the correlation between the private key and the runtime of the cryptographic operation. Consider a smart card that stores a private RSA key. Since the card is tamper resistant, Eve (Attacker) may not be able to examine its contents and expose the key. However, a cleaver attack due to Kocher [99] showed that by precisely measuring the time taken by the smart card to perform an RSA decryption (or signature) Eve can quickly discover the private decryption exponent d.

9.3.4 RSA in Practice

9.3.4.1 Recommended Size of Modulus

In order to foil the powerful quadratic sieve and number field sieve factoring algorithms, a modulus n of at least 768 bits was recommended. For long term security, 1024-bit or larger moduli should be used.

9.3.4.2 Selecting Primes

The primes p and q should be selected so that factoring $n = pq$ is computationally infeasible. In order to avoid the elliptic curve factoring algorithm, p and q should be of the same bit length and sufficiently large. For example, if a 1024-bit modulus n is used, then each p and q should be about 512 bits in length. Another restriction on the primes p and q is that the difference $p - q$ should not be too small. If $p - q$ is small, then $p \approx q$ and hence $p \approx n/2$. Thus, n could be factored efficiently simply by trial division by all odd integers close to $n/2$. If p and q are chosen at random, then $p - q$

will be approximately large with overwhelming probability. Some attacks on RSA are also given for small prime difference.

In addition to these restrictions, many authors recommend that p and q should be strong primes. A prime p is said to be strong prime if the following three conditions are satisfied.

1. $p - 1$ has a large prime factor, say r;

2. $p + 1$ has a large prime factor; and

3. $r - 1$ has a large prime factor.

The reason for condition 1 is to foil Pollard's $p - 1$ factoring algorithm which is efficient only if n has a prime factor p such that $p - 1$ is smooth. Condition 2 foils $p + 1$ factoring algorithm, which is efficient only if n has a prime factor p such that $p + 1$ is smooth. Finally, the condition 3 ensures that the cycling attacks will fail.

9.3.4.3 Choice of e and d

We must choose the encryption and decryption key e and d so that security and efficiency both are maintained. The so-called low encryption exponent attack and low decryption attack should not be possible. A recommended and popular value of e is $2^{16} + 1$ and the decryption key must be as large as n.

9.3.5 Efficiency of RSA

The RSA encryption as well as decryption requires one modular exponentiation. To improve the RSA encryption, we wish to choose small encryption exponent e. Due to low encryption exponent attack, we must take care to choose the encryption exponent e. To prevent from low decryption exponent attack, the decryption exponent must be as large as n. If RSA modulus n is a k-bit number, then typically, d is also k-bit number. Suppose, $k/2$ bits are 1. Using the fast exponentiation technique, decryption requires k squaring and $k/2$ multiplication modulo n. If the RSA modulus is a 1024-bit number, these are 1024 squaring and 512 multiplications. Due to large decryption exponent size, the RSA decryption is very slow. (it is quite slow in comparison to DES, in particular if a smart card is used for decryption, that is, RSA is around 1000 time slower than DES). Using Chinese Remainder Theorem (CRT), we can speedup the RSA decryption process.

9.3.5.1 RSA with CRT

In 1982, Quisquater et al. [133] gave a fast deciphering algorithm for RSA, based on Chinese Remainder Theorem, which speeds up the decryption process to the standard RSA algorithm upto four times. This scheme is sometimes known as QCRSA scheme or RSA-CRT. The decryption process of RSA-CRT works as follows:

Bob decrypts the ciphertext, C, with his private RSA key, d, as follows:

1. He computes $M_p = C^{d_p}$ (mod p), $M_q = C^{d_q}$ (mod q), where $d_p = d$ (mod $p-1$) and $d_q = d$ (mod $q-1$).

2. He computes an integer $M \in \{0, 1, 2, \ldots, n-1\}$ such that $M = M_p$ (mod p), $M = M_q$ (mod q). This M is the plaintext that was encrypted.

3. To find M, he uses the Extended Euclidean algorithm to find y_p and y_q, such that $y_p p + y_q q = 1$, then $M = (M_p y_q q + M_q y_p p)$ (mod n).

Note that, the coefficients $y_p p$ (mod n) and $y_q q$ (mod n) are independent of the ciphertext. They can be pre-computed.

If RSA modulus is $n = pq$, $|p| = |q| = 1/2|n|$. Time to compute c^d (mod n) = $O(log^3 n)$. Time to compute $m_p = c^{d_p}$ (mod p) is $O(log^3 p) = O(log^3 n/2^3)$. Time to compute $m_q = c^{d_q}$ (mod q) is $O(log^3 q) = O(log^3 n/2^3)$. Total time in RSA-CRT is $O(log^3 n/2^2) = O(1/4(log^3 n))$. Thus, the theoretical speedup of RSA-CRT over standard RSA is 4 times.

Example 125 *Consider the Example 119, where $p = 23$ and $q = 31$, $n = pq = 713$ and $\phi(n) = 660$. The encryption key $e = 19$ and the decryption key $d = 139$. Consider the ciphertext $C = 525$, to decrypt 525 using CRT, Bob computes*

$$d_p = d \quad (\text{mod } p-1) = 139 \quad (\text{mod } 22) = 7$$

and

$$d_q = d \quad (\text{mod } q-1) = 139 \quad (\text{mod } 30) = 19$$

$y_p = -4$ *and* $y_q = 3$

$$m_p = c^{d_p} \quad (\text{mod } p) = 525^7 \quad (\text{mod } 23) = 15$$

and

$$m_q = c^{d_q} \quad (\text{mod } q) = 525^{19} \quad (\text{mod } 31) = 15$$

Common solution is $m = 15$.
    In fact,

$$m = (m_p y_q q + m_q y_p p) \quad (\text{mod } n) = (15 \times 3 \times 31 - 15 \times 4 \times 23) \quad (\text{mod } 713) = 15.$$

9.3.6 Semantic Security of RSA

In Chapter 1, we have already discussed the three goals of an adversary. In general, an adversary either tries to find the secret key or to decrypt the target ciphertext (or to get some specific information of the plaintext). In addition to these, adversary may try to distinguish encryption of two plaintexts with success probability exceeding $1/2$. Cryptosystems in which the adversary cannot distinguish ciphertexts in polynomial time, provided that certain computational assumption hold, are said to have *semantic security*. We can define the semantic security as follows:

Definition 9.9 Ciphertext Ditinguishability

Instance: An encryption function $f : X \rightarrow Y$; two plaintexts $x_0, x_1 \in X$; and a ciphertext $y = f(x_i)$, $i \in \{0, 1\}$.

Question: Is i = 1?

If we can answer this question with success probability greater than $1/2$, then cryptosystem is not semantically secure, otherwise it is semantically secure.

The standard RSA is not semantically secure as described below:

1. Given any two plaintexts $m_0, m_1 \in Z_n^*$ and a ciphertext $C = m_i^e \pmod{n}$ where $i \in \{0, 1\}$, we can easily determine whether $m_1^e \pmod{n} = C$ or $m_0^e \pmod{n} = C$.

2. Since $\gcd(e, \phi(n)) = 1$, using the property of Jacobi symbol, we have $\left(\dfrac{c}{n}\right) = \left(\dfrac{c}{n}\right)^e = \left(\dfrac{m}{n}\right)$. Thus, for the given ciphertext, without decrypting, we can determine the Jacobi symbol of the corresponding plaintext. So, RSA encryption leaks some partial information concerning the plaintext.

9.3.6.1   Secure RSA (RSA-OAEP)

One of the basic security criteria for a public key cryptosystem is that the system should be randomized. The concept of Optimal Asymmetric Encryption Padding/Protocol (OAEP) was introduced by Bellare and Rogaway in 1994 [16] to achieve this criteria. It is a padding scheme often used together with RSA encryption. OAEP was subsequently standardized in PKCS # 1. The OAEP algorithm is a form of Feistel network which uses a pair of random oracles G and H to process the plaintext prior to asymmetric encryption. When combined with any secure trapdoor one-way permutation, this processing is proved in the random oracle model to result in a combined scheme which is semantically secure under chosen plaintext attack (IND-CPA). When implemented with certain trapdoor permutations (for example, RSA), OAEP is also proved to be secure against chosen ciphertext attack. There are basically two objectives in introducing OAEP. The first objective is to add an element of randomness which can be used to convert a deterministic encryption scheme (for example, traditional RSA) into a randomized scheme. Second, to prevent partial decryption of ciphertexts (or other information leakage) by ensuring that an adversary cannot recover any portion of the plaintext without being able to invert the trapdoor one-way permutation. In principle, the OAEP works as follows.

Let the security parameter be k (maximum running time that a realistic attacker is able to use is considerably smaller than 2^k). Let b be the RSA modulus size. For security $b \geq 1024$. Set $l = b - k - 1$. Two random oracle functions G and H are modelled as, $G : \{0, 1\}^k \rightarrow \{0, 1\}^l$ is an expansion function, $H : \{0, 1\}^l \rightarrow \{0, 1\}^k$ is a compression function. These two functions are publicly known. The plaintext space is $\{0, 1\}^l$.

1. *Enryption*: To encrypt any plaintext $x \in \{0,1\}^l$, choose $r \in_r \{0,1\}^k$, then ciphertext is

$$y = ((x \oplus G(r)) || (r \oplus H(x \oplus G(r))))^e \pmod{n}.$$

2. *Decryption*: In order to decrypt, receiver computes

$$((x \oplus G(r)) || (r \oplus H(x \oplus G(r)))) = y^d \pmod{n}.$$

Then, he can determine

$$r = r \oplus H(x \oplus G(r)) \oplus H(x \oplus G(r))$$

and then

$$x = (x \oplus G(r)) \oplus G(r).$$

Thus, the plaintext x is randomized as $x \oplus G(r)$ and the random number r is masked as $r \oplus H(x \oplus G(r))$. If G and H are random functions, this cryptosystem is secure against CPA provided that inverting the RSA function $x \to x^e \pmod{n}$ is infeasible. In practice, G and H are constructed from cryptographic hash function. In general, if we improve the security, then either space or speed is compromised. RSA-OAEP also has a drawback that is data expansion. The l bits of plaintexts are encrypted to form $k + l$ bits of ciphertext.

9.4 Rabin Cryptosystem

The security of well-known RSA public key cryptosystem is based on the difficulty of factoring large integer n, which is a product of two large primes p and q. If an efficient algorithm for factoring exists, then the attacker can break the RSA scheme easily. Yet it is not known whether there is some easier way to break RSA other than factoring or not. In 1979, Rabin [136] proposed for the first time another public key cryptosystem and digital signature based on the quadratic residue theory. It is called Rabin Cryptosystem. This scheme is proved to be as secure as factoring. In other words, as long as factorization of large integer into primes remain practically intractable, this scheme remains computationally secure. Theoretically, its security is better than RSA, but it is susceptible to the chosen ciphertext attack. The Rabin Cryptosystem is described as follows.

Key Generation: Alice chooses two large primes p and q such that $p \equiv q \equiv 3 \pmod{4}$ and then computes $n = pq$. The public key for Alice is n and the secret keys are (p,q). The plaintext and ciphertext spaces are $\mathbb{Z}_n^*$.

Encryption: To encrypt any plaintext m, the sender Bob (say) computes $c = m^2 \pmod{n}$ and sends the ciphertext c to Alice.

Decryption: Alice who knows secret keys p and q can compute the square root of c as follows. Alice first computes

$$m_p = c^{(p+1)/4} \pmod{p} \quad \text{and} \quad m_q = c^{(q+1)/4} \pmod{q}.$$

Then by using the Chinese Remainder Theorem on pairs (m_p, m_q), $(-m_p, m_q)$, $(m_p, -m_q)$, $(-m_p, -m_q)$, Alice can compute all the possible four roots of C under modulo n. This is similar to the RSA decryption with CRT. Using Extended Euclidean Algorithm, Alice determines $y_p, y_q \in \mathbb{Z}$ such that $y_p p + y_q q = 1$. She computes

$$c_1 = (y_p p m_q + y_q q m_p) \pmod{n} \quad \text{and} \quad c_2 = (y_p p m_q - y_q q m_p) \pmod{n}.$$

Then, the four square roots of $c \pmod{n}$ are $\pm c_1 \pmod{n}$ and $\pm c_2 \pmod{n}$.

The decryption works because

$$m^2 = c \pmod{pq}$$
$$\Rightarrow m^2 = c \pmod{p} \quad \text{and} \quad m^2 = c \pmod{q}$$
$$\Rightarrow m^{p-1} = c^{(p-1)/2} = 1 \pmod{p} \quad \text{and} \quad m^{q-1} = c^{(q-1)/2} = 1 \pmod{q}.$$

Now

$$m_p^2 = c^{(\frac{p+1}{4})^2} = c^{\frac{p+1}{2}} = c \cdot c^{\frac{p-1}{2}} \pmod{p} = c \pmod{p},$$

and similarly,

$$m_q^2 = c \pmod{q}.$$

Thus, m_p and m_q are square root of c under modulo p and q respectively.

Remark 9.17 If $p \equiv q \equiv 3 \pmod{4}$ then, we can compute the square root of c $\pmod{n}$ efficiently as described above. If $p \equiv q \equiv 3 \pmod{4}$, then $\frac{p+1}{4}$ is an integer, so, $c^{\frac{p+1}{4}}$ is an integer power of c. If $p \equiv q \equiv 1 \pmod{4}$ then, we can compute the square root but the computation becomes more difficult.

Example 126 *Bob selects $p = 23$ and $q = 7$ (both are congruent to 3 mod 4). Bob calculates $n = pq = 161$. Bob announces n publicly and keeps p and q private. Alice wants to send the plaintext $m = 24$. Note that, 161 and 24 are relatively prime so, 24 is in Z_{161}^*.*

Encryption: *She calculates $c = 24^2 = 93 \pmod{161}$ and sends the ciphertext 93 to Bob.*

Decryption: *Bob receives 93 and calculates the four values*

$$m_p = 93^{\frac{23+1}{4}} \pmod{23} = 1 \pmod{23}, \qquad -m_p = 22,$$
$$m_q = 93^{\frac{7+1}{4}} \pmod{7} = 4 \pmod{7}, \qquad -m_q = 3.$$

Using CRT on pairs $(1,4),(1,3),(22,4),(22,3)$, *Bob finds four possible plaintexts:* 116, 24, 137, *and* 45.

Note that, only the second root is the actual plaintext.

An advantage of Rabin cryptosystem is that it is a provable secure cryptosystem however, it has some disadvantages in practice, such as the ambiguity of four plaintexts to one ciphertext. But, the problem of ambiguity can be avoided by the knowledge of side information of the plaintext. For example, the plaintext must be in English words yet when the random key is transmitted, there is no such knowledge to distinguish the four solutions. In practice, this problem can be overcome by adding pre-specified redundancy to the original plaintext before encryption (for example, the character of the message, that is, even or oddness and the Jacobi symbol of the message or last 64 bits of the plaintext). Then, with a high probability, exactly one of the four square roots of a legitimate ciphertext c has this redundancy. So, the receiver can select this one as the intended plaintext. But, then Rabin Cryptosystem is no more provably secure and the ciphertext becomes longer than RSA.

9.4.1   *Efficiency of Rabin Cryptosystem*

We can compare the computational efficiency of RSA and Rabin cryptosystem. In RSA encryption, one exponential power e modulo n is required, whereas the Rabin encryption only requires one squaring. So the Rabin encryption, is more efficient than RSA with the smallest possible encryption key $e = 3$ in RSA. Decryption in the Rabin system is as expansive as RSA decryption with CRT. It requires one exponentiation modulo p and one exponentiation modulo q and one application of CRT.

9.4.2   *Cryptanalysis of Rabin Cryptosystem*

In this section, we show that Rabin Cryptosystem is provably as secure as factoring against ciphertext only attack but is completely insecure under CCA.

9.4.2.1   *Security against Ciphertext Only Attack*

If there is an efficient algorithm that can factor the modulus n, applying that algorithm one can decrypt any ciphertext and get the corresponding plaintext. Conversely, suppose, the attacker Oscar can break the Rabin system. Let n be the public Rabin modulus and p, q its prime factors. Let $\mathcal{R}$ be the algorithm that breaks Rabin system. That is, given a square element C in $Z_n^\star$, it computes m in $Z_n^\star$ such that $m = \mathcal{R}(c)$ and $m^2 = C \pmod{n}$. Now we explain how algorithm $\mathcal{R}$ can be used to factor n. Oscar chooses at random an integer $x \in \{0, 1, 2 \dots, n-1\}$. If $\gcd(x, n) = 1$; then this gcd is equal to one of the prime factor of n. Hence, the factorization of n is found. Otherwise, Oscar computes $C = x^2 \pmod{n}$ and $m = \mathcal{R}(C)$. Here, m is a square root of C

(mod n). It is not necessarily equal to x but m satisfies one of the four congruences:

$$m = x \pmod{p} \quad \text{and} \quad m = x \pmod{q} \qquad \text{(i)}$$
$$m = -x \pmod{p} \quad \text{and} \quad m = -x \pmod{q} \qquad \text{(ii)}$$
$$m = x \pmod{p} \quad \text{and} \quad m = -x \pmod{q} \qquad \text{(iii)}$$
$$m = -x \pmod{p} \quad \text{and} \quad m = x \pmod{q} \qquad \text{(iv)}$$

In case (i), we have $m = x$, and hence, $\gcd(m-x,n) = n$. In case (ii), we have $m = n - x$, and hence, $\gcd(m-x,n) = 1$. In case (iii), we have $\gcd(m-x,n) = p$. In case (iv), we have $\gcd(m-x,n) = q$.

Since x has been chosen at random with equidistribution, each of those cases have the same probability.

Therefore, this procedure factors n with probability at least $1/2$. After k applications of this procedure, n is factored with probability at least $1 - (1/2)^k$. Thus, increasing the reparation, we can factor the modulus with high success probability.

9.4.2.2 Security of Rabin against CCA

If an attacker can obtain a decryption assistance to decrypt any chosen ciphertext, the decryption assistance plays exactly the same role as the algorithm used to find square root in the proof against ciphertext only attack. Thus, the system is completely insecure against CCA.

Let us consider an example to illustrate the above attack.

Example 127 *For $n = 253$. Suppose, Oscar is able to compute square root modulo 253 with algorithm $\mathcal{R}$. He chooses $x = 17$ and obtains $\gcd(17,253) = 1$. Then he computes $C = 17^2 \pmod{253} = 36$. The square root of 36 modulo 253 are 6, 17, 236, and 247. Now $\gcd(6 - 17, 253) = 11$, $\gcd(247 - 17, 253) = 23$. If $\mathcal{R}$ yields one of those square roots, then attacker has found the factorization.*

9.5   ElGamal Cryptosystem

Besides RSA and Rabin, another public key cryptosystem is ElGamal, proposed by T. ElGamal in 1985 [58]. ElGamal is based on the discrete logarithm problem. The details of ElGamal cryptosystem are as follows.

Key Generation: Choose a prime number p, construct the corresponding cyclic group $\mathbb{Z}_p^\star$. Choose a primitive root $g \pmod{p}$ and an integer $a \in_u \{2,3,4\ldots,p-2\}$. Compute $A = g^a \pmod{p}$. Public key of the user (Bob, say) is (p,g,A) and the private key is a.

Encryption: Let the plaintext be $M \in \mathbb{Z}_p^\star$. To encrypt the message M, sender (Alice, say) first chooses a random number $r \in \{2,3,\ldots,p-2\}$ and computes $C_1 = g^r \pmod{p}$ and $C_2 = MA^r \pmod{p}$. The complete ciphertext is the pair (C_1, C_2).

Decryption: To decrypt the ciphertext (C_1, C_2), Bob first computes $k = C_1^a = g^{ar}$ (mod p) and then obtains the plaintext m as $m = k^{-1}C_2$ (mod p).

9.5.1 Correctness of Decryption

Note that, the decryption works correctly since

$$C_2 k^{-1} = MA^r g^{-ar} \pmod{p} = Mg^{ar} g^{-ar} \pmod{p} = M \pmod{p}.$$

Remark 9.18 To avoid the calculation of multiplicative inverse, instead of using $M = C_2(C_1^a)^{-1}$ (mod p) for decryption, we can use Fermat's Little Theorem to compute the plaintext as $M = C_2 \times C_1^{p-1-a}$ (mod p).

Example 128 *Bob first chooses $p = 11$ and $g = 2$, and $a = 3$. Computes $A = 2^3$ (mod 11) = 8. So, the public key for Bob is $(2, 8, 11)$ and the private key is 3.*

Encryption: *To encrypt a plaintext $M = 7$, Alice chooses $r = 4$ and calculates C_1 and C_2 as follows,*

$$C_1 = 2^4 \pmod{11} = 16 \pmod{11} = 5$$

and

$$C_2 = MA^r \pmod{p} = 7 \times 8^4 \pmod{11} = 7 \times 4096 \pmod{11} = 6 \pmod{11}.$$

Alice sends the complete ciphertext $(5, 6)$ to Bob.

Decryption: *Bob computes the plaintext M as follows,*

$$(5^3)^{-1} \times 6 \pmod{11} = 3 \times 6 \pmod{11} = 7 \pmod{11}.$$

9.5.2 Efficiency

ElGamal decryption, like RSA decryption, requires one modular exponentiation (one inversion or multiplication is also required which can be neglected). Unlike RSA, the CRT, however, does not speed up ElGamal decryption. RSA encryption requires only one modular exponentiation, whereas the ElGamal encryption requires two modular exponentiation and one multiplication. The exponentiation in encryption are independent of the plaintext that is actually encrypted. Those exponentiation can be carried out as pre-computations. The actual encryption requires only one modular multiplication and is therefore much more efficient than RSA encryption. For the security, the pre-computed values must be kept secret and must be securely stored, such as on a smart card.

In ElGamal cryptosystem, the ciphertext is twice longer than the plaintext. This is called message expansion and is a disadvantage of this cryptosystem. However, ElGamal cryptosystem is randomized cryptosystem which can be regarded as an advantage. Due to this, ElGamal cryptosystem is semantically secure, for some special primes, whereas RSA and Rabin are not.

9.5.3 ElGamal and Diffie-Hellman

Statement: Breaking ElGamal cryptosystem is not harder than computing DLP.

If there exists an efficient algorithm $\mathcal{A}$ to solve DLP, using the algorithm we can compute the secret key a of the user in ElGamal and hence, can compute the plaintext for the target ciphertext. However, it is still an open problem that if we can break the ElGamal cryptosystem then can we solve the DLP?

Statement: Breaking ElGamal cryptosystem and breaking Diffie-Hellman key exchange protocol are equally difficult.

Suppose there is an attacker, Oscar, having algorithm $\mathcal{A}$ which can solve Diffie-Hellman Problem, that is, the algorithm $\mathcal{A}$ which on input (p, g, g^a, g^r) outputs g^{ar} (mod p). Oscar wants to decrypt ElGamal ciphertext (C_1, C_2). He knows the public key (g, A, p). Since he can break Diffie-Hellman system, he determines the key $k = g^{ra}$ (mod p). He then constructs the message $m = k^{-1}C_2$ (mod p). Conversely, suppose Oscar can break the ElGamal Cryptosystem, that is, he has an algorithm $\mathcal{B}$ which on input (C_1, C_2, p, g, A), outputs the plaintext m, where $A = g^a$ (mod p), $C_1 = g^r$ (mod p) and $C_2 = m.A^r$ (mod p). To find the DH shared key, Oscar chooses $C_2 = 1$ and runs the algorithm for the input $(C_1, C_2 = 1, p, g, A)$ and gets the plaintext m. Clearly, m satisfies the relation $1 = m.A^r$ (mod p) which implies that $m = k^{-1}$ (mod p), the inverse of shared DH key.

9.5.4 Semantic Security of ElGamal

ElGamal cryptosystem is a randomized cryptosystem. It seems to be semantically secure. But, the basic ElGamal cryptosystem is not semantically secure. Recall, in ElGamal Cryptosystem $g \in Z_p^*$ is a primitive element, $A = g^a$ (mod p) where a is the private key. The encryption is defined as $E_k(m, r) = (C_1, C_2) = (g^r, mA^r)$ (mod p). Now, by Euler's criterion, A is QR modulo p if and only if a is even (that is, LSB of a is 0) and C_1 is QR modulo p if and only if r is even. Using quadratic residuesity test, (Euler's criterion) we can test the QR modulo p of A and C_1. Thus, we can determine the parity (LSB) of both a and r and hence, can compute the parity (LSB) of ar. Using the LSB of ar, we can determine if $A^r (= g^{ra})$ is QR modulo p or not.

Now suppose for given a ciphertext (C_1, C_2), we want to distinguish encryption of two plaintexts m_1 and m_2, where $m_1 \in QR_p$ and $m_2 \in QNR_p$ (that is, m_1 is quadratic residue modulo p and m_2 is quadratic non-residue modulo p). Since $m_1 \in QR_p$, so $C_2 \in QR_p$ if $A^r \in QR_p$. Thus, it is clear that (C_1, C_2) is encryption of m_1 if and only if A^r and C_2 are both in QR_p or both in QNR_p. Thus, testing the quadratic residuecity of A^r and C_2, we can determine whether the plaintext m_1 corresponds to the given ciphertext or not.

The above attack does not work if $A = g^a$ (mod p) is in QR_p and every plaintext also belongs to QR_p. If we choose the prime p as $p = 2q + 1$, where q is also a prime, and if we choose g as a primitive qth root of 1 modulo p (we choose g such that $ord(g) = q$, that is, $1 = g^{\frac{p-1}{2}}$ (mod p), clearly $g = g_1^2$ (mod p), where g_1 is a generator of Z_p^*). Then the qth root of 1 is always a quadratic residue element in Z_p^*. In this case A, and C_1 are always QR elements. If both the plaintexts are also QR

elements then it is infeasible to distinguish the encryption of these two plaintexts from the given ciphertext. This version of ElGamal Cryptosystem is conjectured to be semantically secure if the DLP in $Z_p^\star$ is intractable.

9.5.5 Malleablity of ElGamal Cryptosystem

ElGamal is Malleable. Malleable means "attacker can change the ciphertext in such a way that the plaintext is changed in a controlled way". Consider (C_1, C_2) to be a ciphertext that corresponds to message m. Then $(C_1, C_2x \pmod{p})$ is the ciphertext that corresponds to plaintext mx for any $x \in \{0, 1, 2 \ldots, p-1\}$. Thus, for a known pair of plaintext and ciphertext, attacker can modify the ciphertext as per his desired modification on plaintext.

9.6 Elliptic Curve Cryptosystem

Elliptic curve is one of the oldest and the most fascinating concepts of algebraic geometry. It has been extensively studied by mathematicians for many years. It had yielded significant results in the later half of the 20th century. Elliptic curve cryptography is just one application of elliptic curve theory. The use of elliptic curve in cryptography began when Lenstra [102] discovered a factorization algorithm over it. There after, Koblitz [95] and Miller [119] independently proposed the concept of elliptic curve cryptosystem (ECC) by adapting the existing cryptographic protocols on elliptic curves. The main advantage of ECC is that it provides the same security level as RSA and others, with smaller key size. The theory of elliptic curves is quite difficult and beyond the scope of this book. We will only introduce the rudiments that will be useful for our purpose. One can refer the books [159, 21] for further study of elliptic curve. Elliptic curve can be defined over real numbers, any ring, and finite field. Here, we describe the elliptic curve over a finite field.

Definition 9.10   Let C be a curve and P be a point (x, y) on C. Then P is a singular point on C if both the partial derivatives vanish at P. A curve with at least one singular point is a singular curve; otherwise it is a non-singular curve.

Definition 9.11   An affine Weierstrass equation C over a field F is an equation of the form
$$C : y^2 + a_1xy + a_3y = x^3 + a_2x^2 + a_4x + a_6$$
where $a_1, a_2, a_3, a_4, a_6 \in F$.

Definition 9.12   An elliptic curve E is a curve defined by a non-singular Weierstrass equation
$$E : y^2 + a_1xy + a_3y = x^3 + a_2x^2 + a_4x + a_6 \tag{9.5}$$

where the following two equations $a_1 y = 3x^2 + 2a_2 x + a_4$, $2y + a_1 x + a_3 = 0$ cannot be satisfied simultaneously by any point (x, y) on the curve E.

9.6.0.1 Elliptic Curve over a Field F_p

The elliptic curve can be defined over any field and ring, but elliptic curve over finite field is of practical importance in cryptography.

Definition 9.13 Let F be a finite field of characteristic $\neq 2, 3$ and let $x^3 + ax + b = 0$ (where $a, b \in F$) be a cubic with no multiple roots. An elliptic curve $E(a, b)$ over F is the set of points $(x, y) \in F \times F$ satisfying the Weierstrass equation

$$y^2 = x^3 + ax + b \qquad (9.6)$$

together with a single element denoted by ∞ and called the point at infinity.

The left-hand side of Equation 9.6 has a degree of 2 while the right-hand side has a degree of 3. This means that a horizontal line can intersect the curve in three points if all roots are real. However, a vertical line can intersect the curve at most in two points.

Remark 9.19 Using a suitable transformation, we can reduce the general Weierstrass Equation 9.5 into Equation 9.6 if characteristic of the field F is not equal to 2 and 3.

Remark 9.20 The discriminant of equation $x^3 + ax + b = 0$ is equal to $(\frac{a}{3})^3 + (\frac{b}{2})^2$. If $4a^3 + 27b^2 \neq 0$, then the curve is called non-singular elliptic curve. In this case $x^3 + ax + b = 0$ has three distinct roots. If $4a^3 + 27b^2 = 0$, the equation is called singular elliptic curve.

Remark 9.21 If F is the prime field F_p, then the equation $x^3 + ax + b = 0$ has no multiple roots if and only if $\gcd(4a^3 + 27b^2, p) = 1$.

The set of points on elliptic curve form an Abelian group under some binary operation "+" as defined below.

9.6.1 Addition Operation of Elliptic Curve $E(a, b)$ over F_p (Chord and Tangent Method)

Assume that $4a^3 + 27b^2 \neq 0 \pmod{p}$. Define,

$$E_p(a, b) = \{(x, y) : y^2 = x^3 + ax + b \pmod{p}, 0 \leq x, y < p\} \bigcup \{\infty\}. \qquad (9.7)$$

Then $E_p(a, b)$ is an elliptic curve and forms an Abelian group under the binary operation "+" as follows.

1. $P + (-P) = \infty$, $P + \infty = \infty + P = P$; ∞ is the identity element with respect to addition.

2. If $P = (x_1, y_1) \in E_p(a, b)$, then $-P = (x_1, -y_1)$.

3. If $Q = (x_2, y_2) \in E_p(a, b)$ and $Q \neq -P$ then, $P + Q = (x_3, y_3)$ is defined by

$$x_3 = \lambda^2 - x_1 - x_2, y_3 = \lambda(x_1 - x_3) - y_1$$

and

$$\lambda = \begin{cases} \frac{y_2 - y_1}{x_2 - x_1} & \text{if } (x_1, y_1) \neq (x_2, y_2), \\ \frac{3x_1^2 + a}{2y_1} & \text{if } (x_1, y_1) = (x_2, y_2). \end{cases}$$

9.6.1.1 Geometrical Explanation of the Addition Operation

Suppose, we want to add the points $P = (x_1, y_1)$ and $Q = (x_2, y_2)$ on the elliptic curve $E_p(a, b)$ as defined in Equation 9.7. Let l be the line connecting P and Q (tangent line if $P = Q$) and let R be the third point of intersection of l with $E(a, b)$. If l' is the line connecting R and ∞, then $P + Q = R'$ is the point such that l' intersects $E_p(a, b)$ at R, and $P + Q$, that is, $P + Q$ is the symmetric image of R (see Figure 9.1 and Figure 9.2). With the composition "+", $E_p(a, b)$ forms an Abelian group with identity element ∞. We can calculate the formula for the addition operation "+" as given below.

Let the line connecting P and Q be $l : y = \lambda x + c$. Explicitly, the y-intercept and the slope of l are given by $c = y_1 - \lambda x_1$ and

$$\lambda = \begin{cases} \frac{y_2 - y_1}{x_2 - x_1}, & \text{if } (x_1, y_1) \neq (x_2, y_2), \\ \frac{3x_1^2 + a}{2y_1} & \text{if } (x_1, y_1) = (x_2, y_2). \end{cases}$$

Now, we find the intersection of $E_p(a, b) : y^2 = x^3 + ax + b$, $a, b \in F_p$, and $l : y = \lambda x + c$ by solving $(\lambda x + c)^2 = x^3 + ax + b$. We already know that x_1 and x_2 are solutions, so we can find the third solution x_3 by comparing the two sides of $x^3 + ax + b - (\lambda x + c)^2 = (x - x_1)(x - x_2)(x - x_3)$. Equating the coefficients of x^2, gives $\lambda^2 = x_1 + x_2 + x_3$ and hence $x_3 = \lambda^2 - x_1 - x_2$. Then we compute y_3 using $y_3 = mx_3 + c$ and finally $P + Q = (x_3, -y_3)$.

In Short: Addition algorithm for $P = (x_1, y_1)$ and $Q = (x_2, y_2)$ on the elliptic curve E is:

1. If $P \neq Q$ and $x_1 = x_2$ then $P + Q = \infty$.

2. If $P = Q$ and $y_1 = 0$ then $P + Q = 2P = \infty$.

3. If $P \neq Q$ (and $x_1 \neq x_2$), then define $\lambda = \frac{y_2 - y_1}{x_2 - x_1}$ and $c = \frac{y_1 x_2 - y_2 x_1}{x_2 - x_1}$.

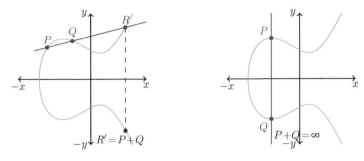

Figure 9.1: Elliptic Curve Addition

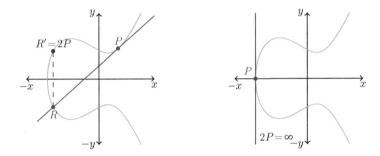

Figure 9.2: Point Doubling in Elliptic Curve

4. If $P = Q$ and $y_1 \neq 0$, then define $m = \frac{3x_1^2 + a}{2y_1}$ and $c = \frac{-x^3 + ax + b}{2y}$

Then $P + Q = (\lambda^2 - x_1 - x_2, \lambda(x_1 - x_3) - y_1)$.

Remark 9.22   The point at infinity ∞ is actually in the form of (x, ∞). Any line joining (x, y) and $(x, -y)$ will intercept at infinity, that is, at (x, ∞) and its inverse image is also (x, ∞) (considering negative of infinity as infinity).

Definition 9.14   An elliptic curve $E_p(a, b)$ is said to be supersingular elliptic curve if $p | t$, where order of elliptic curve group (defined by $\#E_p(a, b)) = p + 1 - t$, p is a prime. Otherwise, it is called non-supersingular elliptic curve.

Definition 9.15   Let G be a cyclic subgroup of order n of $E_p(a, b)$ and P be the generator of G, that is, $nP = \infty$. For any given $Q \in G$ and P, to compute x such that $Q = xP$ is called elliptic curve discrete logarithm problem (ECDLP).

Remark 9.23 For supersingular elliptic curves, the reduction of ECDLP in elliptic curve group G to DLP in finite field F_p is possible in polynomial time (the MOV attack [116]).

Remark 9.24 Elements of finite field F_p are integers between 0 and $p-1$. The prime number p is chosen such that there is finitely large number of points on the elliptic curve to make the cryptosystem secure. US Security and exchange commission (SEC) specifies curves with p ranging between 112-521 bits.

To find the number of point on any elliptic curve is also very difficult. But Hasse's theorem bounds the number of points to a restricted interval.

Theorem 9.3 Hasse's Theorem

The order of the elliptic curve $E_p(a,b)$ over F_p is given by $\#E_p(a,b) = p+1-a_p$, where $|a_p| \leq 2\sqrt{p}$.

For any $k \in F_p$ the scalar multiplication operation $\times$ is defined below:

$$k \times (x,y) = \overbrace{(x,y) + (x,y) + (x,y) + \cdots + (x,y)}$$

k times over $E_p(a,b)$.

Add-and-Double Method for Point Multiplication

Similar to the square and multiply method, we use "Add-and-Double Method" in elliptic curve multiplication.

For example, to compute $18P$, we do 4 doubling and one addition $18P = 2P + (2(2(2(2P))))$.

Example 129 *Consider the equation $y^2 = x^3 + 2x + 3$ over the field F_{17}, the calculation is done under modulo 17. The points on elliptic curve are given below:*

P	$-P$	P	$-P$
$(2,7)$	$(2,10)$	$(11,8)$	$(11,9)$
$(3,6)$	$(3,11)$	$(12,2)$	$(12,15)$
$(5,6)$	$(4,11)$	$(13,4)$	$(13,13)$
$(8,2)$	$(8,15)$	$(14,2)$	$(14,15)$
$(9,6)$	$(9,11)$	$(15,5)$	$(15,12)$

Clearly, $|E(F_{17})| = 20$. Let us add two points $P+Q = R$, where $P = (2,7)$ and $Q = (3,6)$ as follows.

- $\lambda = (6-7)(3-2)^{-1} \pmod{17} = 16$

- $x = (16^2 - 2 - 3) \pmod{17} = 13$.

- $y = 16(2-13) - 7 \pmod{17} = 4$.

- $R = (13,4)$, *which is in the given curve.*

Let $2P = (x_2, y_2)$, *where* $P = (2,7)$.

- $\lambda = (3 \times 2^2 + 2)(2 \times 7)^{-1} \pmod{17} = 1$

- $x_2 = (1^2 - 2 - 2) \pmod{17} = 14$

- $y_2 = 1(2 - 14) - 7) \pmod{17} = 2$

- $2P = (14, 2)$

9.6.2 Elliptic Curves over $GF(2^n)$

To define an elliptic curve over $GF(2^n)$, one needs to change the conventional equation. The common equation is

$$y^2 + xy = x^3 + ax^2 + b \tag{9.8}$$

9.6.2.1 Addition Law of Elliptic Curve $E(a,b)$ over F_{2^n}

The addition operation "+" of elliptic curve over the field F_{2^n} is defined as follows:

1. ∞ is the identity element with respect to addition; $P + (-P) = \infty$, $P + \infty = \infty + P = P$.

2. If $P = (x_1, y_1) \in E(a,b)$, then $-P = (x_1, x_1 + y_1)$.

3. If $Q = (x_2, y_2) \in E(a,b)$ and $Q \neq \pm P$, then $P + Q = (x_3, y_3)$ is defined by

$$x_3 = \lambda^2 + \lambda + x_1 + x_2 + a, y_3 = \lambda(x_1 + x_3) + x_3 + y_1$$

where $\lambda = \frac{y_2 + y_1}{x_2 + x_1}$.

4. If $Q = P$, then $R = P + Q = (x_3, y_3)$ is defined by

$$x_3 = \lambda^2 + \lambda + x_1 + a, y_3 = x_1^2 + (\lambda + 1)x_3$$

where, $\lambda = x_1 + \frac{y_1}{x_1}$.

Here, the elements of the finite field are integers of length at most m bits. In a binary polynomial, the coefficients can only be 0 or 1. The m is chosen such that there are finitely large number of points on the elliptic curve to make the cryptosystem secure. The elliptic curve with m ranging between 113 to 571 bits is recommended to use for secure cryptosystem.

Example 130 *We choose $GF(2^3)$ with elements $0, 1, g, g^2, g^3, g^4, g^5, g^6$ using the irreducible polynomial of $f(x) = x^3 + x^2 + 1$, which means that $g^3 + g^2 + 1 = 0$ or*

$g^3 = g^2 + 1$. *Other powers of g can be calculated accordingly. See the following table.*

0	000	$g^3 = g^2 + 1$	101
1	001	$g^4 = g^2 + g + 1$	111
g	010	$g^5 = g + 1$	011
g^2	100	$g^6 = g^2 + g$	110

Consider the elliptic curve $y^2 + xy = x^3 + x + g$, (here $a = 1, b = g$). Then, the points of elliptic curve will be

$$\left\{ (0, g^4), (g^2, 0), (g^2, g^2), (g^3, g), (g^3, g^4), (g^4, 1), (g^4, g^6) \right\}.$$

Let us add two points $P = (0, g^4)$ and $Q = (g^2, 0)$. In this case, $\lambda = (0 + g^4)(g^2 + 0)^{-1} = g^2$, $x_3 = g^4 + g^2 + 0 + g^2 + 1 = g^4 + 1 = g^2 + g = g^6$ and $y_3 = g^2(0 + g^2 + g) + g^2 + g + g^4 = g + 1 = g^5$. Thus, $P + Q = (g^6, g^5)$.

Let us find $R = 2P = (x_3, y_3)$ for $P = (g^2, g^2)$. In this case, $\lambda = g^2 + \frac{g^2}{g^2} = g^2 + 1 = g^3$, $x_3 = g^6 + g^2 + 1 + 1 = g$ and $y_3 = g^4 + (g^2 + 1 + 1)g = g^4 + g^3 = g^2$. Thus, $R = (g, g^2)$.

Example 131 *Consider the elliptic curve, $E_{17}(2, 2) = \{(x, y) : y^2 = x^3 + 2x + 2 \pmod{17}\}$ and a point $P = (5, 1)$. To find the cyclic subgroup $G = \{Q \in E_p(2, 2) : Q = rP, 0 < r < \#P\} \cup \{\infty\}$ of $E_{17}(2, 2)$ generated by P, we keep adding P to itself to compute $2P, 3P, \ldots$. We first compute $2P = P + P = (x_3, y_3)$:*

$$\lambda = \frac{3x_1^2 + a}{2y_1} = (2 \cdot 1)^{-1}(3 \cdot 5^2 + 2) = 2^{-1} \cdot 9 = 9 \cdot 9 = 13 \qquad \pmod{17};$$

$$x_3 = \lambda^2 - x_1 - x_2 = 13^2 - 5 - 5 = 159 = 6 \qquad \pmod{17}; \ and$$

$$y_3 = \lambda(x_1 - x_3) - y_1 = 13(5 - 6) - 1 = -14 = 3 \qquad \pmod{17}.$$

Thus, $2P = (5, 1) + (5, 1) = (6, 3)$. Similarly, we get all the other points of G to get

$$G = \begin{cases} P = (5, 1), & 2P = (6, 3), & 3P = (10, 6), & 4P = (3, 1), & 5P = (9, 16), \\ 6P = (16, 13), & 7P = (0, 6), & 8P = (13, 7), & 9P = (7, 6), & 10P = (7, 11), \\ 11P = (13, 10), & 12P = (0, 11), & 13P = (16, 4), & 14P = (9, 1), & 15P = (3, 16), \\ 16P = (10, 11), & 17P = (6, 14), & 18P = (5, 16), & 19P = \infty \end{cases}.$$

In this example, the order of P is $n = 19$. Hence, G is a cyclic group of order $\#G = 19$.

9.6.3 Elliptic Curve Cryptosystem

The cryptosystems based on DLP and factoring problem over $\mathbb{Z}_n^*$ can be generalized to elliptic curve, for example, ECDHP, ECElGamal, ECRSA, ECDSA etc. ECC based on DLP is more useful than the ECC based on factoring.

9.6.3.1 Elliptic Curve DH Protocol (ECDHP)

Let G be a cyclic subgroup of an elliptic curve $E_p(a,b)$ of order n. Let $P = (x,y)$ be a generator of G. Suppose, Alice and Bob want to share a secret key $K \in E_p(a,b)$. They proceed as follows.

1. Alice chooses $d_A \in \mathbb{Z}_n^\star$ and computes $K_1 = d_A P \in E_p(a,b)$. Keeps d_A as secret key and sends K_1 to Bob using any channel.

2. Bob chooses $d_B \in \mathbb{Z}_n^\star$ and computes $K_2 = d_B P \in E_p(a,b)$. Keeps d_B as secret key and sends K_2 to Alice using any channel.

3. After receiving K_1, Bob computes $K = d_B K_1$ in $E_p(a,b)$.

4. After receiving K_2, Alice computes $K' = d_A K_2$ in $E_p(a,b)$.

5. Clearly, $K = K' = d_A d_B P$ is the common shared secret key.

The security of the above ECDHP is based on elliptic curve discrete logarithm problem (ECDLP).

Example 132 *Let Alice and Bob agree on $p = 17$ and the elliptic curve as defined in Example 131 with generator $P = (5,1)$.*

■ *Alice chooses $2 \in \mathbb{Z}_{19}^\star$ and computes $K_1 = 2P = (6,3)$. Sends $(6,3)$ to Bob and keeps 2 as her secret key.*

■ *Bob chooses $7 \in \mathbb{Z}_{19}^\star$ and computes $K_2 = 7P = (0,6)$. Sends $(0,6)$ to Alice and keeps 7 as his secret key.*

■ *Bob computes the shared key $7(6,3) = (9,1)$.*

■ *Alice computes the shared key $2(0,6) = (9,1)$.*

■ *Thus, the shared key between Alice and Bob is $(9,1)$.*

9.6.3.2 Elliptic Curve based ElGamal Cryptosystem

Suppose, Bob wants to generate keys for ECElGamal Cryptosystem . He proceeds as follows.

Key Generation: 1. Bob chooses an elliptic curve $E_p(a,b)$ over F_p.

2. Bob chooses a point $P = (x_1,y_1) \in E_p(a,b)$ of order n (say) (hence, $nP = \infty$).

3. Bob randomly chooses $d \in \mathbb{Z}_n$

4. Bob computes $Q = d \times P = (x_2,y_2)$.

5. Bob announces $E_p(a,b), P, Q$ as his public key and keeps d as his secret key.

Encryption: Suppose, Alice wants to send a plaintext $m = (x, y) \in E_p(a, b)$ to Bob. She first chooses a random integer r and computes $C_1 = rP$ and $C_2 = M + rQ$. Sends the ciphertext (C_1, C_2) to Bob.

Decryption: After receiving the ciphertext (C_1, C_2), Bob first computes $C = dC_1 = rdP$ and then computes the plaintext $m = C_2 - C$.

Example 133 *We consider the same elliptic curve defined in Example 131.*

Key Generation: *Bob chooses $d = 3$ and computes $3P = (10, 6)$. Bob's public key is $E_{17}(2, 2), P = (5, 1), Q = (10, 6)$ and his secret key is $d = 3$.*

Encryption: *Suppose Alice wants to send a message $(7, 11) \in E_{17}(2, 2)$ to Bob. She chooses $r = 5$ and computes $C_1 = 5P = (9, 16)$ and $C_2 = (7, 11) + 5(10, 6) = (16, 13)$. Sends (C_1, C_2) to Bob.*

Decryption: *Bob first computes $C = dC_1 = 3(9, 16) = (3, 16)$ and then computes $m = C_2 - C = (16, 13) - (3, 16) = (7, 11)$.*

9.6.3.3   Advantages and Disadvantages of ECC over RSA

We can compare ECC with the most popular and practical PKC, RSA. The advantages and disadvantages of ECC over RSA are discussed below.

Advantages

The main advantage of ECC is that it provides the same security level with smaller key size. Due to smaller key size, the encryption process becomes faster and requires less computing power than RSA. For example, the ECC with 160-bit encryption key provides the same security as RSA with 1024-bit key and depending upon the implemented platform ECC can be around 15 times faster than RSA. Both RSA and ECC are practically useful. The ECC is more applicable in wireless devices where battery life, computing power, and memory are limited.

Disadvantages

One of the main disadvantages of ECC over RSA is message expansion. The plaintext and ciphetext size in RSA are same but in ECC, the ciphertext size is twice of plaintext's size. Thus, the size of ciphertext in ECC is significantly more than RSA. Another disadvantage of ECC over RSA is that the algorithm in ECC is more complex and more difficult to implement.

By Table 9.1, it is obvious that smaller parameters can be used in ECC than in RSA and DL based systems at the same security levels. As discussed above, the advantages of the smaller parameters are that they provide speed (they are faster in computation), smaller key sizes, and certificates. In this way, the private key operations (signature generation or decryption) and the public key operations (verification or encryption) are much more efficient in ECC than RSA and DL based cryptography. Moreover, the advantages of ECC can be realized in the systems where processing power, storage, bandwidth, or power consumption are crucial.

Table 9.1: Comparison of key sizes on equivalent level of securities

Security (in bits) →	80 (SKIPJACK)	112 (Triple-DES)	128 (AES-128)
DL/EC parameter	160	224	256
RSA/DL modulus	1024	2048	3072

9.7 Exercises

1. Generate two 8-bit prime numbers p and q such that the RSA modulus $n = pq$ is a 16-bit number and the public RSA key $e = 3$ can be used. Compute the corresponding private key d. Encrypt the string 110100110110100 with the public exponent 5.

2. Alice encrypts m with Bob's public RSA key $(3869, 47)$. The ciphertext is 3266. Determine the plaintext m.

3. How many operations are required for an RSA encryption with encryption exponent $2^{16} + 1$?

4. The message m is encrypted by the RSA system using the public keys $(391, 5)$ and $(391, 3)$. The ciphertexts are 195 and 304. Use the common modulus attack to find m.

5. The same message m is encrypted by the RSA system using the public keys $(391, 3), (319, 3)$, and $(205, 3)$. The ciphertext are $340, 201$, and 59. Use low-exponent attack to find m.

6. Alice encrypts m with Bob's public RSA key $(403, 7)$. The ciphertext is 291. Determine the plaintext m.

7. Suppose, there are two users having common RSA modulus and relatively prime encryption keys. If the same plaintext m is encrypted for both the users, then the plaintext can be recovered from the corresponding two ciphertexts. How?

8. Suppose that RSA modulus is $n = 403$, the encryption exponent is $e = 7$, and decryption exponent is $d = 103$. Describe how to factor the modulus n with the help of e and d.

9. Give an example of adaptive chosen ciphertext attack on RSA.

10. Alice receives the ElGamal ciphertext $(C_1 = 30, C_2 = 40)$. Her public key is $(p = 41, g = 7)$ and the secret key is $a = 8$. Determine the corresponding plaintext.

11. Let $p = 3539, g = 7, A = 37$ be Bob's public ElGamal key. Let the plaintext message is GERMANS ARE COMMING. Converting into numbers using $A = 00, B = 01, \ldots, Z = 25, 1 = 26, 2 = 27, \ldots, 9 = 34, space = 35$. Encrypt the entire plaintext using the random number $r = 5$.

12. Let $N = 77$ be a Rabin modulus and let the ciphertext $c = 23$ be obtained using Rabin encryption algorithm. Determine all the possible plaintext.

13. Let the RSA modulus $n = 403$ and the decryption key $d = 103$. Explain how to decrypt a ciphertext $C = 219$ using CRT.

14. Suppose, that three users in a network, say Ajay, Vijay, and Sanjay, all have public RSA encryption exponents $b = 3$. Let their RSA moduli be denoted by n_1, n_2, n_3 and assume that n_1, n_2 and n_3 are pair-wise relatively prime. Now suppose Rani encrypts the same plaintext x send to Ajay, Vijay, and Sanjay using RSA cryptosystem. Describe how an attacker Eve can compute x, given $y_1, y_2,$ and y_3 without factoring any of the moduli.

15. Alice and Bob share a secret key $k = g^{ab} \pmod{p}$ using Diffe-Hellman protocol with $p = 467$, $g = 4$, $a = 400$ and $b = 134$. Later, they shared another secret key k_1 with the same p, q, b and with $a_1 = 167$. At the end of the protocol they found that $k = k_1$. Determine the value of both shared keys and explain why the keys are identical.

16. Let (n, e) be a public key for an RSA cryptosystem. Suppose, c_1 is the ciphertext that corresponds to the plaintext m, and that c_2 is the ciphertext if the plaintext $(m + 1)$ is encrypted. Show that if $e = 3$, then m can be found by solving the congruence equation $(c_2 - c_1 + 2)x = c_2 + 2c_1 - 1 \pmod{n}$.

17. Prove that the factoring RSA modulus is computationally equivalent to computing secret RSA key from the public RSA key.

18. Explain how to generate a third ElGamal ciphertext of an unknown plaintext with the help of two ElGamal ciphertexts. How can this attack be prevented?

19. Show that in the RSA cryptosystem the decryption exponent d can be chosen such that $ed = 1 \pmod{lcm(p - 1, q - 1)}$.

20. To simplify the implementation of DH protocol one replaces the multiplication group $(Z_p^\star, \cdot)$ by the additive group $(Z_p, +)$. How is the security effected?

21. Discuss the semantic security of RSA and ElGamal cryptosystems.

22. Prove that the Rabin Cryptosystem is provably as secure as factoring against ciphertext only attacks.

23. Prove that the factoring RSA modulus is computationally equivalent to computing the secret RSA key from the public RSA key.

24. Why is man-in-the-middle attack possible in DH protocol? How to secure DH protocol from this attack?

25. How to prevent RSA from Pollard $p-1$ factoring algorithm?

26. Find all the points of elliptic curve $y^2 = x^3 + 3x + 2$ over the field F_{13}. Using addition rule, find the result of the addition of two different points and same points over the curve.

27. Find all the points in the the elliptic curve $y^2 + xy = x^2 + x + g^2$ over the $GF(2^4)$ using the irreducible polynomial $f(x) = x^4 + x + 1$. Also, discuss addition of two points in the elliptic curve.

Chapter 10

Digital Signature

CONTENTS

Security is one of the prime concerns in cryptography. The integral building blocks of security have been discussed in Chapter 1. To recall, they are privacy/confidentiality, authentication, integrity, and non-repudiation. Digital signature is a cryptographic primitive which can help to achieve authentication, integrity and non-repudiation. A digital signature scheme is a part of asymmetric cryptosystem which offers method of signing a message using private key and its verification using public key. A digital signature can be regarded as an electronic value, a number, or a binary string; which depends on private key of the signer and content of the message. A digital signature must be verifiable, and in case of a dispute like whether the signature is made by the legitimate signer or not, an unbiased third party should be able to determine the consequence without requiring access to the signer's private key. Besides authentication, integrity, and non-repudiation, digital signature has a significant application in the certification of the public keys in large networks.

10.1   Formal Definitions

We start the chapter with formal definitions of basic elements involved in digital signature.

10.1.1   Basic Elements of Digital Signature

Basically, a digital signature scheme can be described by the two algorithms:

- **Signature:** This takes as input a message and private key, and returns a signature on the message.

- **Verification:** This takes as input the message, signature and public key, and returns true or false, depending on whether the signature was valid corresponding to the message or not.

10.1.2   Formal Structure

Formally, a digital signature can be defined by the five-tuple $(\mathcal{P}, \sigma, \mathcal{K}, \mathcal{S}, \mathcal{V})$ [162], where:

1. $\mathcal{P}$ is a finite set of possible *messages*.

2. σ is a finite set of possible *signatures*.

3. $\mathcal{K}$ is a finite set of possible *keys*, we call it the *keyspace*.

4. For each $k \in \mathcal{K}$, there is a *signing algorithm sign$_k$* $\in \mathcal{S}$ and a corresponding *verification algorithm ver$_k$* $\in \mathcal{V}$.

For consistency, the functions $sign_k : \mathcal{P} \rightarrow \sigma$ and $ver_k : (\mathcal{P} \times \sigma) \rightarrow \{true, false\}$ should be defined in a way that:

$$ver_k(x,y) = \text{true if } y = sign_k(x),$$

for every message $x \in \mathcal{P}$ and for every signature $y \in \sigma$.

10.1.3   Digital Signature Scheme

Usually, as a cryptographic design, the digital signature schemes are formally structured by the following algorithms:

1. $params \leftarrow Setup(\lambda)$: An algorithm which takes as input a security parameter λ and outputs the public parameters *params*.

2. $(sk, vk) \leftarrow KeyGen(params)$: An algorithm which takes as input public parameters *params* and outputs the signing (private) key *sk* and the verification (public) key *vk*.

3. $\sigma \leftarrow Signature(m, sk, params)$: This is usually a probabilistic algorithm run by the signer that takes as input the message *m* to be signed, signing key *sk*, and the public parameter *params* to generate a signature σ.

4. $b \leftarrow Verification(params, vk, m, \sigma)$: A deterministic algorithm run by the verifier that takes as input the public parameter *params*, signer's public key *vk*, the message *m*, and signature σ, and returns a bit *b* which is 1 if the signature is valid and 0 if invalid.

Before discussing other details, we prefer to address following two common issues regarding digital signature:

1. What is the difference between a digital signature and a conventional signature?

   ■ The conventional signature is included in the document; it is a part of the document, whereas digital signature is not included in the document and has to be sent as a separate document.

   ■ Digital signature of a message is intimately connected with the **message**, and for **different messages** it is **different** (one to one relation), whereas the handwritten signature is not dependent to the message but only adjoined to the message and always remains **same** for different messages (one to many relation).

2. Why do we need digital signature?

   ■ We want some signature like function for electronic communications to achieve source authentication, data-integrity, and non-repudiation.

   ■ For example, to use the internet as a safe and secure medium for e-Governance and e-Commerce.

10.2 Attack Goals of an Adversary of a Digital Signature

In cryptography, we study the attack goals of an adversary of a digital signature in the following views.

First, we consider an adversary $\mathcal{A}'s$ view (or attempt) of attack [115]:

■ *Total break*: Either $\mathcal{A}$ is able to compute the private key of the signer or $\mathcal{A}$ can derive an efficient algorithm equivalent to the signing algorithm.

■ *Universal forgery*: $\mathcal{A}$ attempts to forge a digital signature without having the private key of signer.

■ *Selective forgery*: $\mathcal{A}$ can generate a valid signature (for a given public key) for a particular message chosen prior to the attack. This is also known as *target message forgery*.

■ *Existential forgery*: $\mathcal{A}$ can forge a signature for at least one message. In other words, $\mathcal{A}$ can output at least one (m, σ) pair, such that σ is a valid signature on message m, corresponding to the given public key.

These attacks can be classified in the following categories, depending upon the access provided to the adversaries:

■ *Key only attack*: In this attack, an adversary knows only the signer's public key. This attack is also known as *public key only attack* or *passive attack* .

■ *Message attack*: Here, an adversary is given access to the signing oracle corresponding to certain messages. The message attack can be further divided into three classes:

  ■ *Known-message attack*: An adversary can obtain a valid signature for a message known to him.

  ■ *Chosen-message attack*: An adversary can generate valid signatures, for the messages, from a chosen list of messages.

  ■ *Adaptive chosen-message attack*: An adversary can generate valid signatures for the chosen messages depending upon the previously obtained (message, signature) pairs.

After the proposal of identity (ID)-based cryptography [152], digital signatures are designed and studied extensively on ID-based setting. For ID-based setting, we also study the following attacks, in addition to the above:

■ *Chosen-ID attack*: An adversary can obtain valid signatures for a chosen list of identities.

■ *Adaptive chosen-ID attack*: An adversary is given access to the signing oracle for any identity chosen adaptively.

In both the cases, it is understood that the identities submitted to the signing oracle are (ofcourse) other than the target identity on which the adversary wants to forge the signature.

The above definitions imply that a *secure* signature should be robust against the adaptive chosen-message (and adaptive chosen-ID) attack.

10.3 Digital Signature in Practice

Today, digital signature is one of the most useful primitives used for authentication. The first international standard for digital signature (ISO/IEC 9796) was adopted in 1991. This was the RSA signature (signature based on the RSA parameters). Later, in 1993, the Digital Signature Algorithm (DSA) was adopted as the Digital Signature Standard (DSS) by the National Institute of Standard and Technology (NIST) (U.S. Government), which is a variant of the basic ElGamal signature. In the recent literature, the elliptic curve variant of the DSA (known as the elliptic curve digital signature algorithm (ECDSA)) has been studied extensively. Below, we briefly discuss these signature schemes. In the below discussions, we assume that Alice is the signer who sends signed messages to the verifier Bob.

10.3.1 RSA Signature

The RSA signature scheme is based on the parameters of the RSA cryptosystem. Like the RSA cryptosystem, security of the RSA signature scheme is based on the difficulty of solving integer factoring problem. Below we describe the RSA signature scheme.

■ **Setup**: Alice selects two random large primes p and q and computes $N = pq$. Alice also selects a hash function $h : \{0,1\}^\star \to Z_N^\star$.

■ **KeyGen**: Alice selects an integer $v \in Z_N^\star$ such that $v < \phi(N)$ and $gcd(v, \phi(N)) = 1$ (where $\phi(N)$ is Euler's totient function). She also computes integer $s \in Z_N^\star$ such that $sv \equiv 1 \ (\bmod \ \phi(N))$. Finally, Alice publishes her public key or verification key as (v, N). She keeps her private key or signing key s secret with herself.

■ **Signature**: To sign a message m, (essentially $m \in Z_N^\star$, if hash function is not used), Alice computes:

■ $h = H(m)$
■ $\sigma = h^s (mod \ N)$.

Alice sends the signature σ along with the message m to Bob.

■ *Verification*: To verify the signature, Bob computes:

■ $\sigma^v (mod\ N) = h_1$.

■ $H(m) = h_2$

and accepts the signature σ if and only if $h_1 = h_2$.

Remark 10.1 Z_N^* is set of all the integers less than N and relatively prime to N.

Remark 10.2 As discussed in chapter 8, in practice, the user always signs the message digest (i.e. hash image $h(m)$ of the message) and not the original message, otherwise some basic attacks are possible.

Example 134 *See the following example:*
KeyGen: Alice selects $p = 7$, $q = 11$, $v = 17$, obtains $N = 77$ and $s = 53$ (using the Extended Euclidian algorithm). Hence, public key of Alice is $(77, 17)$ and private key is 53. To sign a message $m \in Z_N^*$ with $h(m) = 13$ (say), Alice does the following:

Signature: She computes $\sigma = 13^{53} = 41\ (mod\ 77)$. Alice sends the signature 41 along with the message to Bob.

Verification: Receiving the signature and message pair $(41, m)$, Bob computes $41^{17} = 13 (mod\ 77)$ and validates the signature if $h(m) = 13$.

10.3.1.1 Security of RSA Signature

The security parameters of RSA setup have already been discussed in Chapter 9. The same parameters are applicable for the RSA signature as well. It is also emphasized that for practical purposes we need to pay additional attention to security. One precaution is that RSA should be used with a secure randomized padding scheme. The theoretical RSA discussed in the textbook is often referred as *textbook RSA* to distinguish with the practical RSA.

Why to use hash function in RSA signature? The straight forward attack on RSA is to compute the private key s, which can be done by solving the factoring problem. To avoid the feasibility of such attacks, one simple idea is to choose both p and q large integers of equal size (bitlength). For 1024 bit public key N (as considered to be a secure parameter), p and q are usually considered to be 512 bit in size.

But this is not the only way to forge the RSA signature. There may be other possibilities to forge RSA signature when the signer signs a message without using hash function (i.e. signature is $\sigma = m^s (mod\ N)$). Below, we discuss a few possibilities of attacks on RSA signature without hash function:

1. **No message attack/Existential forgery:** Attacker selects a random $\sigma \in (0, 1, 2, .., N - 1)$ and claims that σ is Alice's signature.

Bob computes $m = \sigma^v \pmod{N}$ and believes that Alice would have signed the message m if it's meaningful.

2. **Known message attack:** If the attacker has two message signature pairs (m_1, σ_1) and (m_2, σ_2), then he can create signature on the third message $m_3 = m_1.m_2$, s.t. $m_3^s = (m_1^s \bmod N)(m_2^s \bmod N)$.

3. **Chosen message attack:**

   (a) Suppose, attacker wants to forge a signature on a message m. For this, he selects m_1 different than m and with $gcd(m_1, N) = 1$, then he computes $m_2 = m \times m_1^{-1} \pmod{N}$ and obtains signature σ_1 and σ_2 on messages m_1 and m_2. Further he gets signature σ on message m as $\sigma = \sigma_1 \times \sigma_2 \pmod{N} = m_1^s.m_2^s = (m_1.m_2)^s \pmod{N}$.

   (b) Alternatively, to forge a signature on message m, the attacker asks Alice to sign a message $m' = r^v \times m$, where r is a randomly selected integer. Alice's signature on m' is $\sigma' = m'^s = rm^s \pmod{N}$. Finally, the attacker sets $\sigma = \frac{\sigma'}{r} \pmod{N} = m^s \pmod{N}$ which is a valid signature.

Remark 10.3 RSA signature without hash function is not secure against known message attack and chosen message attack.

Remark 10.4 To avoid the above forgeries, especially the existential forgery and forgery by known message attack, the message padding with some redundancy is used during signature. When verifier verifies the signature and recovers the plaintext out of the signature, he analyzes the output with the inherent redundancy and accepts only if it consists of the redundant information.

10.3.1.2 Performance

The RSA signing process is an exponentiation with the private key and the verification process is an exponentiation with the public key. Hence, for efficient verification a small number can be chosen for the public key.

10.3.2 ElGamal Signature

The ElGamal signature scheme is based on the parameters of the ElGamal cryptosystem. Like the ElGamal cryptosystem, security of the ElGamal signature scheme is based on the computation of discrete logarithm over the finite field.

- **Setup:** Alice randomly selects a sufficiently large prime p and a primitive root $g (mod\ p)$ of Z_p^*. Alice also selects a hash function $h : \{0,1\}^* \rightarrow \{1,2,3...p-1\}$.

- **KeyGen:** Alice further selects a random value $a \in \{2,..,p-2\}$ and computes:

$$A = g^a (mod\ p)$$

The public key of Alice is (p,g,A) and private key is a.

- **Signature:** To sign a message m, Alice randomly selects $k \in \{2,..,p-2\}$ and co-prime to $(p-1)$.

  She further computes

  - $r = g^k\ (mod\ p)$,
  - $s = k^{-1}(h(m) - ar)(mod\ (p-1))$

  where k^{-1} is inverse of $k\ (mod\ (p-1))$.

  Finally, Alice outputs signature $\sigma = (r,s)$ on the message m.

- **Verification:** Receiving the (message, signature) pair (m,σ), Bob first checks whether
$$1 \leq r \leq (p-1).$$

  If the above condition is not satisfied, he rejects the signature. Otherwise, he checks the following
$$A^r r^s \equiv g^{h(m)} (mod\, p)$$

  if the above relation holds, he accepts the signature, otherwise rejects.

it follows as:

$$A^r r^s \equiv g^{ar} g^{kk^{-1}(h(m)-ar)} \equiv g^{h(m)} (mod\ p).$$

Example 135 *Consider an example.*

- **KeyGen:** *Alice selects $p = 547$, $g = 9$, $a = 23$ and computes $A = g^a = 9^{23}\ (mod\ 547) = 81$. Her public key is $(p = 547, g = 9, A = 81)$ and private key is $a = 23$.*

- **Signature:** *Suppose, Alice wants to sign message m with $h(m) = 100$ (say). For this she does the following:*

  1. *First she selects $k = 125$, note that $(125,546) = 1$, hence co-prime to $p-1$.*

2. *Further Alice obtains* $r = g^k = 9^{125} \pmod{547} = 304$. *Also, she computes the inverse of* $k \pmod{p-1 = 546}$ *as* $k^{-1} = 83$.

3. *Further, Alice computes* $s = k^{-1}(h(m) - ar) \pmod{p-1} = 83 \times (100 - 23 \times 304) \pmod{546} = 172$.

4. *Finally, Alice outputs the signature as* $\sigma = (r, s) = (304, 172)$.

■ **Verification**: *To verify the signature* $(304, 172)$, *Bob proceeds as follows.*

1. *Bob computes* $A^r r^s \pmod{p} = 81^{304} \times 304^{172} \pmod{547} = 81$.

2. *He also computes* $g^{h(m)} \pmod{p} = 9^{100} \pmod{547} = 81$.

3. *So, the signature is valid, as* $A^r r^s \equiv g^{h(m)} \pmod{p}$.

Remark 10.5 The signer must be careful to choose a different k uniformly at random for each signature and to be certain that k, or even partial information about k, is not leaked. Otherwise, an attacker may be able to deduce the secret key x with reduced difficulty which is perhaps enough to allow a practical attack. In particular, if two messages are sent using the same value of k and the same key, then an attacker can compute x directly.

10.3.2.1 Security of ElGamal Signature

1. **Choice of p.**
   In such signatures, the prime number p is usually selected such that computation of discrete logarithm in Z_p^* is hard.

2. **Choice of k.**
   For every new signature, a new exponent k must be chosen. This is guaranteed if k is chosen randomly. A pseudorandom generator (PRNG) perfectly works for it. Let's find an answer for the question that why a new k is advised to be selected every time:

   Suppose, signatures s_1 and s_2 on m_1 and m_2 are generated using the same k.

   Then $s_1 - s_2 = k^{-1}(h(m_1) - h(m_2)) \bmod (p-1)$ and hence, k can be determined if $h(m_1) - h(m_2)$ is invertible modulo $(p-1)$.

   Now, once k is known then by the relation.

   $$s_1 = k^{-1}(h(m_1) - ar) \quad \bmod (p-1)$$

   the private key of Alice can be determined as:

   $$a \equiv r^{-1}(h(m_1) - ks_1) \bmod (p-1).$$

3. **Why** $1 \leq r \leq (p-1)$ **?**

If not, then it is possible to generate new signature from the old signature.

Let (r, s) be the signature on message m. Now, to sign a different message m', the attacker computes

$$u = h(m')h(m)^{-1} \quad \text{mod } (p-1)$$

Attacker also computes
$$s' = su \ mod(p-1)$$

and using CRT computes r' satisfying

$$r' \equiv ru \ mod(p-1), \quad r' \equiv r \ mod \ p.$$

The signature on message m' is $(r's')$. Correctness of the signature can be verified as below.

$$A^{r'}(r')^{s'} \equiv A^{ru}r^{su} \equiv g^{u(ar+ks)} \equiv g^{h(m')} \ mod \ p.$$

It can be easily verified that $r' > p - 1$.

4. **Why is hash function important in ElGamal signature?**

If no hash function is used, then existential forgery is possible.
The verification congruence is

$$A^{r}r^{s} \equiv g^{m} \ mod \ p.$$

Attacker selects two integers u, v with $gcd(v, p-1) = 1$. Then he sets

$$r = g^{u}A^{v} \ (mod \ p), s = -rv^{-1} \ mod \ (p-1), m = su \ mod \ (p-1)$$

with the above values verification congruence holds:

$$A^{r}r^{s} \equiv A^{r}g^{su}A^{sv} \equiv A^{r}g^{su}A^{-r} \equiv g^{m} \ (mod \ p)$$

Remark 10.6 This procedure also works if a collision resistance hash function is used. But since the hash function is one-way, it is likely to be impossible for the attacker to find a message m such that the signature generated is the signature of m.

Security Assumption of ElGamal Signature: Security of ElGamal signature is based on the hardness of the discrete logarithm problem, which is discussed in Chapter 9.

10.3.2.2 RSA vs ElGamal Signature

■ **Signature Generation:** In RSA, we need to compute one exponent to the power s mod N. In ElGamal signature, the computations of $r = g^k$ mod p and k^{-1} mod p do not depend on the message, hence, can be pre-computed. Thus, the actual signature generation cost is only two modular multiplications, which is faster than modular exponentiation in RSA.

■ **Signature Verification:** ElGamal scheme requires three modular exponentiations which is considerably more expensive than RSA signature verification.

■ **Signature Size:** For modulus p of size 1024-bits, size of ElGamal signature is 2048 bits, whereas size of RSA siganture is 1024 bits. This is called message expansion. Thus, the size of ElGamal signature is double than the RSA signature.

10.3.3 Digital Signature Algorithm (DSA)

ElGamal signature was proposed in 1985. Later, in 1991 the National Institute of Standards and Technology (NIST) proposed DSA, which can be observed as a variant of ElGamal signature. To use as Digital Signature Standard (DSS), they considered it an official Federal Information Processing Standard- FIPS 186 in 1993.

■ ***Setup*:** Same as ElGamal signature.

■ ***KeyGen*:** Same as ElGamal signature. Here, an additional prime q is also chosen which is a prime factor of $p - 1$. Further, to compute private key, the signer chooses $a < q$, and computes $A = g^a (mod\ p)$. The public key of Alice is (p, q, g, A) and private key is a.

■ ***Signature*:** To sign a message m, signer generates random signing key k, $k < q$ and computes:

    1. $r \equiv (g^k (mod\ p))(mod\ q)$

    2. $s = k^{-1}(h(m) + a \times r)(mod\ q)$

The signature on message m is $\sigma = (r, s)$.

■ ***Verification*:** Receiving the signature $\sigma = (r, s)$ along with the message m, the verifier first verify that $1 \leq r \leq q - 1$ and $1 \leq s \leq q - 1$, and then computes:

    1. $x = s^{-1}(mod\ q)$

    2. $y = (h(m) \times x)(mod\ q)$

    3. $z = r \times x(mod\ q)$

    4. $v = (g^y \times A^z (mod\ p))(mod\ q)$

If $v = r$ then the signature is verified.

Example 136 *Consider an example*

■ **KeyGen:** *Alice chooses $q = 101$, $p = 78q + 1 = 7879$, $g = 170$, and $a = 75$, then she computes $A = g^a mod\ 7879 = 4567$. Hence, Alice's public key is $(p = 7879, q = 101, g = 170, A = 4567)$.*

■ **Signature:** *To sign a message m, for which $h(m) = 22$ (say), Alice proceeds as follows.*

    1. *Alice chooses a random value $k = 50$ and computes $k^{-1} mod\ q = 50^{-1} mod\ 101 = 99$.*

    2. *Then she computes*

$$r = (g^k mod\ p) mod\ q = (170^{50} mod\ 7879) mod\ 101 = 94$$

      and

$$s = k^{-1}(h(m) + a \times r) mod\ q = 99 \times (22 + 75 \times 94) mod\ 101 = 97$$

    3. *The signature on the message digest 22 is $\sigma = (r, s) = (94, 97)$.*

■ **Verification:** *To verify the signature $\sigma = (r, s) = (94, 97)$, Bob computes*

    1. $x = s^{-1}(mod\ q) = 97^{-1} mod\ 101 = 25, y = (h(m) \times x)(mod\ q) = 22 \times 25 mod\ 101 = 45$, *and* $z = r \times x(mod\ q) = 94 \times 25 mod\ 101 = 27$.

    2. $v = (g^y \times A^z (mod\ p))(mod\ q) = (170^{45} \times 4567^{27} mod\ 7879) mod\ 101 = 2518 mod\ 101 = 94 = r$ *i.e* $v = r$, *hence, the signature is valid.*

Remark 10.7   In 1991, the ElGamal signature was modified as *Digital Signature Algorithm* (DSA).

Remark 10.8   The DSA significantly reduces the signature size by working in a subgroup of $F_P^\star$ of a prime order q. This makes it a good choice for practical implementations.

Remark 10.9   The underlying idea is that using index calculus to solve the DLP in a subgroup of $F_P^\star$ is no easier than solving it in $F_P^\star$ by the same method.

Remark 10.10   The DSA was officially published as *Digital Signature Standard* (DSS) in 1993.

10.3.3.1   Security of DSA

As DSA is a variant of ElGamal signature, hence all the security concerns of ElGamal signature are recommended for DSA also. In particular, to avoid the existential forgery, the use of cryptographic hash function (SHA-1) is strictly recommended. Similarily, the value k should be updated every time for each new signature to avoid the potential key guess like ElGamal. Also, like the ElGamal signature, the private key recovery of the signer in DSA directly corresponds to the solution of discrete log problem (DLP) in the subgroup of $Z_p^\star$. But the fact that solution of DLP in a subgroup of $Z_p^\star$ is no easier than its solution in the group $Z_p^\star$, the security of DSA resembles with the security of Elgamal signature.

10.3.3.2   ElGamal vs DSA

The DSA explicitly uses hash function SHA-1 which outputs 160-bit number. The computations in DSA may look complex than that of ElGamal signature as in the signature phase of ElGamal there are more pre-computed values than DSA which make it efficient for repeated signature generation. But more importantly, DSA is efficient in verification than ElGamal. This is crucial because verification can be repeated many times for a signature. On the top of all computations, the signature size of DSA is six times smaller than ElGamal snd hence is six times faster. It is mainly because ElGamal exponents are calculated modulo the prime number p which is at least 768-bit number as per the DLP parameter, whereas DSA exponent depends on the SHA-1 i.e. 160-bit number. It is important to note that National Security Agency (NSA) America and other organisations prefer DSA over the ElGamal. But as the DSA theory has been built up over the building blocks of ElGamal, ElGamal signature has its own classical importance.

10.3.4   Elliptic Curve Digital Signature Algorithm (ECDSA)

We recall that the idea of elliptic curve cryptosystem introduced in mid 80's was formalized by replacing subgroup of $Z_p^\star$ by the group of points on elliptic curve. Security of the systems based on elliptic curve corresponds to the solution of the elliptic curve discrete logarithm problem (ECDLP), defined below. As the ECDLP appears to be harder than DLP (no sub-exponential algorithms are available like index calculus), the smaller parameter serves equal security as achieved by DLP with large parameter. The ECDSA is the elliptic curve analogue of the DSA. It was accepted as an ISO (International Standards Organization) standard `ISO 14888-3` in 1998. In 1999, it was considered as an ANSI (American National Standards Institute) standard `IANSI X9.62`. Further, it was accepted as an IEEE (Institute of Electrical and Electronics Engineers) standard `IEEE 1363-2000`, and a FIPS standard `FIPS 186-2` in the year 2000.

Before going to the signature scheme, let us recall the elliptic curve group. The details of elliptic curve group has been already discussed in Chapter 9.

Definition 10.1   Elliptic curve group: Let p be a prime number and F_p denotes the field of integers modulo p. An elliptic curve E over F_p is defined by an equation of the form $y^2 = x^3 + ax + b$ where $a, b \in F_p$ satisfying $4a^3 + 27b^2 \neq 0 \ (mod\ p)$. A pair (x, y), where $x, y \in F_p$, is a point on the curve E if (x, y) satisfy the equation.

The set of all points on E, denoted by $E(F_p)$ form a group under the addition modulo p and is known as elliptic curve group.

- ■ *Setup:* Same as ElGamal signature.

- ■ *KeyGen:* Alice chooses a finite field F_p, an elliptic curve $E(F_p)$ and a point $G \in E(F_p)$ of a large prime order q.

  She further selects a secret number $1 < s < (q-1)$
  and computes $V = sG \in E(F_p)$
  the tuple $(p, q, E(F_p), G)$ is public value of the algorithm. Alice's public key is V and private key is s.

- ■ *Signature:* To sign a message m, Alice selects a random integer $e \in \{2, .., q - 1\}$
  computes $eG \in E(F_p)$ and
  $s_1 = eG \ (mod\ q)$, $h = h(m)$ and $s_2 = (h + su)e^{-1} \ (mod\ q)$
  where $s_1 = (u, v)$

  The signature on message digest h is the pair (s_1, s_2)

- ■ *Verification:* To verify the signature (s_1, s_2), Bob computes

  $v_1 = hs_2^{-1} \ (mod\ q)$
  $v_2 = us_2^{-1} \ (mod\ q)$ and
  $v_1 G + v_2 V$,
  finally verify whether

$$v_1 G + v_2 V \ (mod\ q) = s_1$$

Correctness of the verification is as below

$$
\begin{aligned}
v_1 G + v_2 V &= hs_2^{-1}G + us_2^{-1}sG \\
&= s_2^{-1}G(h + su) \\
&= s_2^{-1}Ges_2 \\
&= eG \\
&= s_1
\end{aligned}
$$

To get a private key s from the relation $V = sG$, while having $V, G \in E(F_p)$ is equivalent to solving the ECDLP:

Definition 10.2   Elliptic curve discrete logarithm problem (ECDLP): Let E be an elliptic curve defined over a finite field F_p. Let $P \in E(F_p)$ be a point of order q, then for given p, P, q, an equation of elliptic curve E and $Q \in E(F_p)$, to determine an integer $n \in Z_q^*$ such that $Q = nP$ holds, is called ECDLP. nP means n times addition of P in elliptic curve.

Remark 10.11   Till now, no efficient algorithm is known to solve ECDLP.

Example 137 *Consider an example of ECDSA.*

- **KeyGen**: *Consider an elliptic curve $y^2 = x^3 + x + 6$ defined over F_{11}, $p = 11$. Alice chooses $G = (2,7) \in E(F_{11})$ of order $q = 13$, and the secret key $s = 7$, she further computes the verification key $V = sG = (7,2)$. The tuple $(11, 13, E(F_{11}), (2,7))$ is public value of the algorithm. Alice's public key is $V = (7,2)$ and private key is $s = 7$.*

- **Signature**: *Suppose Alice wants to sign a message m, for which $h = h(m) = 4$ (say).*

    1. *Alice selects $e = 3$, and computes $s_1 = eG = 3 \times (2,7) = (8,3)$, and*
    2. *$s_2 = (h + su)e^{-1} \ (mod \ q) = (4 + 7 \times 8) \times 3^{-1} \ (mod \ 13) = 7$.*
    3. *The signature on message m (with $h(m) = 4$) is $(s_1, s_2) = \{(8,3), 7\}$.*

- **Verification**: *To verify signature (s_1, s_2), Bob computes*

    1. *$v_1 = hs_2^{-1} = 4 \times 7^{-1} = 4 \times 2 = 8$ and $v_2 = us_2^{-1} = 8 \times 2 = 3$.*
    2. *The signature is valid as $v_1 G + v_2 V = 8 \times (2,7) + 3 \times (7,2) = (8,3) = s_1$.*

10.3.4.1   Security of ECDSA

We have discussed security notions of digital signature in the view of total break, existential forgeability etc. In the above section, we have discussed that obtaining private key of signer in ECDSA directly corresponds to solution of the well-known ECDLP which is proved to be hard and no polynomial algorithm is known for the solution of ECDLP. Hence, in one sense, ECDSA is believed to be secure till ECDLP is a hard problem. Alternatively, as the usual security concern of a digital signature is to be existentially unforgeable against chosen message attack, to the best of our knowledge, there is no proof of security for such an attack over ECDSA but a slight variant of DSA is proven secure from such an attack [132], hence, the corresponding variant of ECDSA is also secure. The attacks on ECDSA can be studied in following categories:

1. Attacks on ECDLP using following algorithms: Pohlig-Hellman algorithm, Baby-step giant-step algorithm, Pollard's Rho algorithm, Xendi calculus attack. Attacks by sieving (Number Field Sieve, Function Field Sieve) etc.

2. Attacks on corresponding hash function.

3. Hardware attacks (including side-channel attack).

10.3.4.2 Why ECDSA over other Signatures?

1. The most important feature of the ECDSA is that it offers the same level of security as DSA (and RSA signature) with comparatively very smaller bit-sized keys. For example, where the DSA requires at least 1024-bit key to achieve 80-bit security, the ECDSA requires only 160-bit key. The ECDSA can be upto 15 times faster than RSA depending on the platform of implementation. Hence, computations can be executed faster, thus ECDSA becomes ideal for security implementation with low processing power, small storage space, and shorter bandwidth channels. This makes ECDSA a good choice for constrained devices such as pagers, cellular phones smart cards etc.

2. Another importance of ECDSA is that its security relies on the solution of ECDLP. The unavailability of efficient algorithm for solution of ECDLP makes ECDSA a good choice for a robust signature algorithm.

10.3.4.3 Application of ECDSA

Very recently, the key exchange for end-to-end encryption of the messages of Whatsapp (a social messaging application on mobile phone) was achieved using ECDSA. Bitcoin is a cryptocurrency which is replacing physical currency exchange. Security of bitcoin depends on ECDSA. Bitcoin address is a cryptographic hash of an ECDSA public key. The ownership of the account is authenticated by verification of ECDSA generated by the account holder. In Apple's recent *white paper on iOS security*, ECDSA is used for Apple ecosystem. Messages through iMessage are signed with ECDSA also iCloud keychain syncing relies on ECDSA.

10.3.4.4 Issues with ECDSA

One of the main disadvantages of ECDSA is that it increases the signature size significantly than RSA signature. Hence, proves to be less efficient than RSA on the practical platform. Furthermore, the ECDSA algorithm relies on the elliptic curve group where only addition is defined. Therefore, it appears less flexible for different computations than RSA. Due to the above reasons, it is very difficult to efficiently implement ECDSA for practical purpose and that is why ECDSA has not been widely used in real life digital securities, and we still depend on RSA signature for banking and other crucial websites.

10.4 Some Popular Digital Signatures

In the current literature of cryptography, various types of digital signature can be found. Some of the most leading signatures are discussed briefly below:

- *Group signature*: Group signature allows a member of a group to anonymously sign a message on behalf of the group and the signature can be verified by a common public key of the group. There is a group manager, who posses some private key, by which he can expose the identity of signer in case of dispute. It is useful in the scenarios when it is sufficient for a verifier to know that a message was signed by any member of a particular group. One of the important applications of group signature is to achieve anonymity of the signer. The idea of group signature was introduced by Chaum and Van Heyst [44]. For more details regarding the definition and structure of group signature the readers can refer the foundation work by Bellare et al. [14]. A brief survey on group signature is presented in [2].

- *Ring signature*: This signature can be performed by any one from a group of members, where each member posses their own keys. In this signature, it is difficult to determine that whose key in the group has been used to create the signature. It differs from the group signature in the mechanism that firstly, there is no way to revoke anonymity of an individual signer; secondly, any group of users can be used as a group, without an additional setup. It is useful to provide an anonymous signature from an office or organization, without revealing the identity of original signer. The concept of ring signature was first formalized by Rivest, Shamir and Tauman [140] in 2001. For the review of available ring signatures, the survey by Wang et al. [168] can be referred.

- *Proxy signature*: This is a signature generated by an authorized proxy agent of an original signer. In this scheme, a signer may transfer his/her signing rights (without transferring the private key) to any proxy agent who can sign a message on behalf of the signer and the signature can be verified using public keys of both- the original signer and the proxy signer. One or a group of signers may delegate their signing authority to one or any group of proxy signers. Depending upon the number of signers in the original group and the proxy group, the signatures are catagorized in multi-proxy signature (one original signer and many proxy signers), proxy multi-signature (many original signers and one proxy signer), and multi-proxy multi-signature (many original signers and many proxy signers). It is useful in the scenario when the original signer/signers is/are not available to sign the document(s). Applications of proxy signature are suggested in distributed systems, grid computing, mobile agent environments, distributed shared object systems etc. The idea of proxy signature is due to Mambo, Usuda and Okamoto [111]. Detailed survey on proxy signature are presented in [52, 164].

■ *Blind signature*: This is a signature on a message signed by another party (i.e. who has not selected the message or in other words, who has no control over the message) without having any information about the message. Hence, as the name of the signature, the message is blind to the signer. Blind signatures are especially employed in privacy-related protocols where the signer and the message author are different parties. Most significant examples of blind signature are digital cash scheme, electronic voting etc. The concept of blind signature was conceived by David Chaum [43] in 1983. A detailed survey of blind signature can be found in [157].

For review of some pairing-based signature schemes, the survey in [145] can be referred.

10.5 Exercises

1. Show that the RSA signature scheme is not secure against the *known message attack* and the *chosen message attack*.

2. Determine the forgery in ElGamal signature scheme, if the choice of k is not random.

3. How is DSA better than ElGamal Signature?

4. What is the verification congruence in the ElGamal signature scheme if s is computed as $s = (ar + kh(m)) mod\ p$, where $r = g^k mod p$, a is the secret key of signer and k is randomly chosen in $\{2, 3, ..., p - 2\}$?

5. In the DSA signature if a signer has pre-computed a pair (k, r) with $r = (g^k mod\ p)\ mod\ q$ and always uses this pair to sign her messages, show, if it is possible to recover the signer's private key.

Chapter 11

Research Directions in Cryptography

CONTENTS

All the previous chapters of this book deal with the basic introduction to cryptography and their mathematical foundation. However, cryptography is one of the emerging areas of research in mathematics and computer science nowadays. Due to being a widespread area of study, cryptography covers a huge list of different cryptographic schemes. Also, there are numerous areas of research in cryptography and new areas are being explored continuousely. Hence, in this last chapter of the book we flash light on few of the topics which have not been discussed in previous chapters and few of the research topics in cryptography which have attracted researchers in the current years. This chapter covers a very short list of areas which is just the tip of the iceberg. The inclusion of many more such existing topics is beyond the scope of the book.

11.1 Pairing-based Cryptography

In cryptography, pairing or bilinear pairing is essentially a (bilinear) map between two groups. Though the notion of bilinear pairing is quite old, the elementary realization of pairings for cryptographic purpose was due to Menezes et al. [115] for its importance in reduction of elliptic curve logarithms to logarithms in a finite field, popularly known as the MOV attack. The functionality of bilinear pairing has made various decisional problems easier as well (for example decisional bilinear Deffie-Hellman problem [26]) which has given rise to the idea of *gap groups*, where decisional problems are easy but the computational problems are hard. In the definition of the above mentioned bilinear map, the domains are usually these gap groups. The basic importance of pairing in cryptography is to transfer a hard problem in one group (usually the gap group) to a different but usually an easier problem in the other group which offers construction of new cryptographic protocols. The notion of pairing has received attention in cryptography as it offers useful functionalities for practical protocols. This practicality is essentially due to existence of efficient algorithm (e.g. Miller's algorithm [120]) to compute the value of pairing in co-domain.

The popularly known implementations of pairings include the Weil pairing and the Tate pairing which are well-defined mathematical maps. The construction of first fully functioning and practical identity-based encryption [28] is due to the Weil pairing. The application of pairing was also realized in the construction of first fully functioning non-interactive identity-based key agreement scheme [90], one round three party key agreement protocol [89] and short signature [31]. These revolutionary works [28, 31, 89, 90] established strong foundation of pairing in modern cryptography. The application of pairing is majorly recognized in construction of various identity-based cryptosystems due to its unique functionality.

It turns out that elliptic curve serves suitable groups for pairing implementation according to the cryptographic objectives [10, 60]. Hence, the study of elliptic curve is inherently associated with the study of pairings in cryptography. This also offers large range of cryptosystems and their securities based on a variety of elliptic curve. The elliptic curves are specially categorized in the view of pairing implementation viz. supersingular curves [60, 61], non-supersingular curves [10], hyper elliptic curve [48, 56, 78, 96, 156] etc. The descriptions in [107] can be referred for a comparative study between Weil pairing and Tate pairing in the view of different curves. The *pairing friendly curves* [9, 10, 39] and pairing computation [49, 148] have been studied widely in pairing-based cryptography. They can be considered as individual areas of research in pairing. The authors in [149] and [131] have discussed various elliptic curves more precisely with their support to pairing implementation. The language, library and toolkits for software implementation of pairing have been formalized in [7, 8, 53, 106].

Another, comparatively recent, class of study in pairing is *asymmetric pairing*. A pairing $e : G_1 \times G_2 \to G_T$ is called a symmetric or a Type 1 pairing if $G_1 = G_2$, otherwise it is called asymmetric. Asymmetric pairings are further categorized into Type 2 and Type 3 pairings. If there exists an efficiently computable isomorphism between G_1 and G_2, then the pairing is referred to as Type 2, whereas if there is no efficiently computable isomorphism between G_1 and G_2, then the pairing is referred to as Type 3. This classification was initially studied in detail by Galbraith et al. in [62] and by Smart et al. [160] in the view of cryptographic application. An extended type of asymmetric pairing, Type 4 pairing, is discussed in [42] with some implementation notes on pairing and their performance precisely in comparison to the symmetric or Type 1 pairing. Recently, efficient cryptosystems have been designed using asymmetric pairing (specially Type 3 pairing) [41, 45, 137] and claimed to be secure on stronger security notions than previous standards. The Relic toolkit [7] is especially suggested for implementation of asymmetric pairing.

For a review of pairing-based cryptographic protocols, the elementary survey by Dutta et al. [55] can be referred. A recently published book *Guide to Pairing-Based Cryptography* by Mrabet and Joye [57] has enumerated this topic in detail.

11.2   Zero-knowledge Proof System

In the zero-knowledge (ZK) proof technique, the prover tries to prove a second party, the probabilistic polynomial-time verifier, that a given (mathematical) statement is true, in such a way that no additional information needs to be provided for the proof and hence the verifier learns nothing except the only fact that the statement is true. Intuitively, it ensures that all the computations carried out by the verifier can be independent of the protocol (the game between the prover and verifier). One of the important applications of ZK proofs in cryptography is authentication. When a user or sender wants to prove its identity to another user or receiver by the fact that she posses a secret information (private key/decryption key) in such a way that the other party should not learn anything about the decryption key. The notion of ZK proof was first proposed by Goldwasser et al. [74]. ZK proofs are described in [134] with simple examples including the popular magical cave example. Goldreich and Oren [73] classified the zero-knowledge in two categories *auxiliary input zero-knowledge* and *black box simulation* and suggested that the auxiliary input zero-knowledge is more suitable in cryptography than black box simulation [74]. There are two methods in this proof system-interactive and non-interactive. The system established in [74] was interactive. A ZK proof satisfies the properties of completeness, soundness, and zero-knowledge, where the first two properties are in general for any interactive protocols the third property is particularly for the ZK protocols. In [63] authors have formalized that how interactive ZK proofs of knowledge support public key cryptosystems against chosen ciphertext attack. A strong application of ZK proof was demonstrated by Goldreich et al. [72] to construct a ZK proof for any NP-system by any commitment scheme. Various important and basic properties of ZK proof system are discussed in [73]. In a non-interactive ZK (NIZK) proof a common reference string [24], which is usually a set of values used for computation by both the parties prover and verifier, is shared, which enables verification of the proof without requiring actual interaction between the parties. A more functional notion for NIZK proofs, the multi-theorem zero-knowledge proofs, is suggested by Feige et al. [59]. Due to the relevant functionality of the NIZK proof, it is widely used in cryptography for CCA-security of encryption scheme, anonymous authentication, and security of group signature and ring signature. Among the various useful results in NIZK proof technique established by Groth et al. [79, 80, 81, 82] the notions formalized in [81, 82] have been greatly used to conceive a NIZK protocol for various concrete cryptographic schemes, especially for the pairing-based protocols, due to the subgroup hiding technique. For more details and state of art of the topic the extensive reviews in [71, 172] can be referred.

11.3 Authenticated Group Key Exchange

Key exchange or agreement protocol is the first important phase in a secure communication, for example, in a confidential video conferencing among security officials. The popular work of two-party key exchange technique of Diffie and Hellman [54] was a foundation not only for the key exchange protocol but also for the public key cryptography. A key exchange protocol allows the users, communicating over an insecure public channel, to share the common key. As in the real world communication several users have to communicate among them securely, many efforts have been made towards extending the two-party protocol into multi-party protocol. Joux's protocol [89] for three party communication is the first functioning and practical protocol in this direction. Further, it was extended to multi-party protocol in [11]. A group key exchange (GKE) protocol is a cryptographic primitive for exchange of a common key, among a group of multiple members communicating over a public setup, such that the key remains secret to others except the members of the group. Clasically, the security of a GKE protocol lies in its authenticity [36] meaning whether the shared information, which contributes in the computation of the common key, has been communicated from the authentic source or not. Otherwise an attack known as the *man-in-the-middle* (MITM) attack can be applied for impersonation. The other important security aspect is *forward security* which guarantees the security of shared key even after the future corruption. This corruption is referred to be weak [37] if the long-term secret keys are revealed and is referred to be strong [38] if the ephemeral key (session-dependent information) is leaked. The security is also observed in the view of privilege to the attacker i.e. whether an adversary is active or passive. The classical Diffie-Hellman protocol provides security only against passive adversaries. The stronger security requirement is to guarantee authentication and integrity of the communication even against active adversaries.

The provable security of authenticated key exchange (AKE) protocol were introduced in [35, 36, 37]. In [93] authors have advised some crucial improvement over the previous security notions and suggested a construction with possibility of scalability of unauthenticated protocol into the authenticated one. Further, the notion of *insider security* in GKE was proposed in [92]. Another important security property *key compromise impersonation resilience (KCIR)* is discussed in [75] and extended in [179].

In group key agreement (GKA) protocol, the identity-based setting were initially introduced in [46, 138]. These proposals did not satisfy the forward secrecy, later it was achieved in [173, 174]. There are still many objectives to be achieved and hence, the topic is alive, for example, the ephemeral key leakage [179] is yet to be checked and removed in identity-based setting.

11.4 Attribute-based Cryptography

Attribute-based cryptography (ABC) is an extended alternative to identity-based cryptography (IBC) where the attributes of users replace the role of identity. In the IBC, usually the encryptor indicates only one decrypter, in ABC the encrypter can indicate many decrypters by associating the decrypters with certain attributes. The importance of attributes were first considered to provide fine-grained access control over the encrypted data by an attribute-based encryption [77]. Based on the access policy, the attribute-based encryption (ABE) can be observed in two flavours- Key-Policy ABE (KP-ABE) [77] and Ciphertext-Policy ABE (CP-ABE) [19] depending upon the fact that whether *private keys* or the *ciphertexts* are associated with the access structure. As in the KP-ABE, the access policy depends on the user's private key, this may not be suitable in various applications as the data owner then has to trust on the key issuer. Based on the type of access structure, ABE schemes can be categorized as monotonic or non-monotonic, depending upon the fact that whether the private keys can represent a *fixed* access formula or *any* access formula over attributes. The access structure in KP-ABE is a monotonic access structure. The ABE with non-monotic access structure was introduced in [128]. A technique for the transformation of a KP-ABE system into a CP-ABE was initiated in [144]. Further the transformational approach was generalized in [76] using a "universal access tree". Various other extensions of ABE have been explored with different security objectives such as verifiable ABE scheme [163], hierarchical ABE scheme [167] etc.

Attribute-based signature (ABS) is another popular domain of research in ABC. ABS contributes as an anonymous signature in cryptography. The idea of ABS was introduced in [108, 109] as a candidate alternative for signatures in the multi-authority settings. In the current research aspects of ABS, the extended ABS primitives such as attribute-based group signatures [94], attribute-based ring signature [104] etc. are major attarction. The elementary security setups for ABS were realized in the random oracle. Further, since the proposal of [127], the security guidelines were established in standard model which support practical application. The proposal [127] uses dual pairing vector spaces [126] and the functional encryption proof technique of [103]. A threshold variation of the construction is formalized in [150]. The threshold ABS restricts the signer to maintain a threshold number of attributes in common with the verification set of attributes. The first *constant-sized* signature was introduced in [83] with selective-predicate and adaptive-message settings. The security is evidenced in the standard model.

11.5 Homomorphic Encryption

In abstract algebra, homomorphism is defined as a map preserving all the algebraic structures between the domain and range of an algebraic set [110]. If we consider encryption as a function or map e between the set of plaintexts $\mathcal{P}$ and ciphertexts $\mathcal{C}$, then we say that this encryption is a homomorphic encryption if the mapping satisfies the property of *homomorphism* i.e. ciphertext is a homomorphic image of plaintext. Due to this property, the homomorphic encryption allows operations on the encrypted data without affecting the encryption (i.e. it allows a party to operate on the encrypted data without decrypting it). For example, for messages m_1 and m_2, this property of homomorphism can be used to compute directly $e(m_1 \star m_2)$, by just having $e(m_1)$ and $e(m_2)$, where $\star$ is the operation defined in the set of plaintext $\mathcal{P}$. A straightforward example is that RSA and ElGamal encryptions are multiplicatively homomorphic. Due to the property of malleability, homomorphic encryption offers elegant application in cryptography, for example, to address the confidentiality of data in cloud computing environment.

Rivest et al. [141] had used the idea of homomorphism in modern cryptography for the first time for computation without decryption. The popular works in [29, 85, 121] are a few examples where computation on the encrypted data has been performed. The class of homomorphic encryption can be categorized into three types of schemes with respect to the number of allowed operations on the encrypted data (i) *partially homomorphic encryption* (PHE) in which only one type of operation is allowed, however, the operation can be repeated many times (ii) *somewhat homomorphic encryption* (SWHE) where more than one but limited number of operations are allowed with limited number of repetitions (iii) *fully homomorphic encryption* (FHE) where one can perform unlimited number of operations with unlimited number of repeatations. PHE has been observed to be useful in e-voting [17] and Private Information Retrieval (PIR) [101]. But PHE can be used only for some limited applications as it allows only one operation so, either addition or multiplication. In contrast to PHE, SWHE offers choice for more operations (e.g. both addition and multiplication). But, in SWHE, the size of the ciphertexts grows with each homomorphic operation, hence, the total operations are again limited. This limitation gave raise to the desire of a technique which can offer unlimited number of homomorphic operations with random functions. Fully homomorphic encryption (FHE) [67] is a tool which offers this functionality. Though, the first FHE proposed by Gentry [67] is a good alternative to the then existing homomorphic encryption schemes, but as [67] uses lattice-based constructions it is complex and not smooth for practical implementations. However, due to the introduction of lattice into cryptography, many concrete results were established. In particular, Dijk et al. [166], Brakerski and Vaikuntanathan [34] and López et al. [105] have proposed more advanced construction of FHE from lattice and similar assumptions. The second issue in Gentry's seminal work was the key generation algorithm which was intentionally used for a specific purpose only and the generation of lattice from a "good" basis was left without a solution. For the purpose, Gentry introduced a new key generation algorithm in [68]. Later, in [69, 125, 147] the key generation of FHE [68] has been optimized.

For various information about the review of homomorphic encryption we rely on the vast survey of this topic in [1]. There are other good literatures [4, 129, 165] which cover the state of the art of homomorphic encryption and FHE in detail.

11.6 Secure Multi-party Computation

In many applications of electronic transactions, a group of people require to jointly conduct some computation depending on the personal inputs of each party. These computations may take place among mutually trusted parties, partially trusted parties, and untrusted parties. For such computation, one party must usually know the inputs from all the other participants; but if the participants are untrusted then privacy is the top priority. In other words, if a group of mutually untrusted people want to compute some result depending upon inputs of each individual in the group, not revealing the individual inputs to anyone, then the problem is referred to as multiparty computation problem. A solution to achieve this computation securely is called secure multi-party computation (SMPC). The area is also known as secure function evaluation (SFE).

The SMPC basically aims to provide two properties (i) *Input privacy*: this property ensures that the message communicated by a participant does not enable computation of any other additional information, other than the output of the function alone at the last, and (ii) *Correctness*: any proper subset of adversarial colluding parties should not be able to force honest parties to output an incorrect result during the protocol execution. Objective of this property is to ensure that either the honest parties always compute the correct output (this is called a robust MPC protocol) or they abort if they find an error (this is called MPC protocol with abort).

In the last two decades, the topic of SMPC has been widely studied and explored. The notion of multi-party computation was first studied by Yao [177, 178] in the two-party scenario and later extended by Goldreich et al. [70] in the multi-party setting. Their approach is based on secret sharing of all the inputs and zero-knowledge proofs for corrupt parties. Private information retrieval (PIR), privacy-preserving statistical database, and privacy-preserving data mining are some very significant examples of SMPC.

11.7 Secret Sharing

Secret sharing scheme is about sharing a secret in parts among finite number of (authorized) parties. The distribution is done in such a way that certain number of participants from the authorized group can pool their shares to reconstruct the secret and no party outside the authorized group can do this. The distributor of the secret is usually called a dealer in the scheme. Application of secret sharing is realized in building secure authentication protocols with weak passwords, hence, the leakage of a single party's information does not lead to any attack like dictionary attack on the password etc. Secret-sharing schemes are very useful in building box of many secure protocols e.g. generalized protocol for multi-party computation, threshold cryptography, attribute-based encryption, and general oblivious transfer etc.

The secret-sharing schemes were introduced by Shamir [151] and Blakley [22] with the threshold mechanism as discussed above. Secret-sharing schemes for general access structures were introduced and constructed by Ito et al. [86]. The secret sharing scheme has effective application in the real world secure communication. For example, this scheme can secure a secret over multiple servers and remain recoverable despite multiple server failures. For security reasons, most of the modern HSM (hardware secure module) made for cryptographic applications use Shamir's secret sharing scheme [151].

11.8   Post-Quantum Cryptography

Security of almost all well-known cryptosystems are based on some hard mathematical assumptions. For example, security of the well-known RSA system is based on the intractability of integer factorization problem (IFP), security of the ElGamal and related cryptographic schemes are based on the hardness of discrete logarithm problem (DLP) etc. These systems are secure till there are no efficient (i.e. polynomial-time) algorithm to solve these hard problems. By many theoretical and practical observations, it is believed till date that in the available computation resources (e.g. 64 bit computer) it has still not been possible to formalize any algorithm which can solve these hard problems in feasible time (i.e. in polynomial-time). The existing algorithms for the solution of these problems are still exponential (some are sub-exponential too) which would take infeasible time by the available resources for the solution of the problems. Towards cryptanalysis of such cryptosystems, one effort is to devise polynomial-time algorithms which can be suitable for the existing resources to yield a solution in permissible time, the other option is to think of building some very fast computational resource where the computations which take infeasible time in the available machines can be performed within the permissible range of time. The coming era of quantum computers provides this second possibility. More precisely, Prof. Peter W. Shor of MIT has already proposed quantum algorithms for solutions of IFP and DLP in 1994 [158]. So, once the practical quantum computers will be available, then all the existing and popular cryptosystems (based on number theoretic problem) e.g. [58, 95, 142] will be insecure using Shor's algorithm [158]. Taking this threat into account, researchers are already involved in the construction of quantum-resistant cryptosystems, meaning the cryptosystems for which even the quantum computer cannot solve the underlying problem. This wing of research in cryptography is known as *post-quantum cryptography*, which is currently attracting many researchers due to its appealing property of being quantum-immune. The broad families of mainstream post-quantum cryptography are code-based cryptography [114], hash-based cryptography[118], multivariate-quadratic-equations cryptography [130], lattice-based cryptography [84], cryptography from learning with errors [139] and most recently the isogeny-based cryptography [50, 87, 88]. The book

'*Post-Quantum Cryptography*, edited by Bernstein, Buchmann and Dahmanis [18], has several references of the recent works in the post-quantum cryptography.

11.9   Side-Channel Analysis

Security is the prime concern in design of useful cryptosystems. Broadly, there are two ways to analyze the security of an algorithm. First is the theoretical proof through strong mathematical evidences and the other is by practical demonstration through actual implementation. Unfortunately, the system whose security is proved in theory cannot be guaranteed to be secure always during practical implementation. In fact, many times the implementation causes leakage of some additional (side) information during the experiment which can lead to an attack to the system. Such attacks due to these side-channel information are studied in cryptography as side-channel analysis. The appealing feature of these attacks are that they are always or mostly based on small piece of *additional information* and do not require knowledge of the actual algorithm. There are various ways to capture these small but decisive information to approach an attack. Some of the popular side-channel analysis are- *power analysis attack* [97, 98, 112] where the power consumption during the execution of the algorithm is recorded and analyzed to retrieve the secret key, *fault attack* [20] where a faulty value is induced in the algorithm and then the faulty output (due to the induced fault) and actual output are analyzed to recover the secret key, *timing attack* [100] where the time of execution of algorithm is noticed to exploit the secret information, *electromagnetic attack* [3, 135] where the electromagnetic radiations emitted during the execution is captured and analyzed to reveal the secret information, *acoustic attack* [65, 66, 153] where the sound produced during a computation is exploited in order to attack on the scheme.

References

[1] Abbas Acar, Hidayet Aksu, A Selcuk Uluagac, and Mauro Conti. A survey on homomorphic encryption schemes: Theory and implementation. *arXiv preprint arXiv:1704.03578*, 2017.

[2] Aayush Agarwal and Rekha Saraswat. A survey of group signature technique, its applications and attacks. *International Journal of Engineering and Innovative Technology (IJEIT)*, 2(10), 2013.

[3] Dakshi Agrawal, Bruce Archambeault, Josyula R Rao, and Pankaj Rohatgi. Attack strategies-the em side–channel (s). *Lecture Notes in Computer Science*, 2523:29–45, 2002.

[4] S Sobitha Ahila and KL Shunmuganathan. State of art in homomorphic encryption schemes. *International Journal of Engineering Research and Applications*, 4(2):37–43, 2014.

[5] Alfred V Aho and John E Hopcroft. *The design and analysis of computer algorithms*. Pearson Education India, 1974.

[6] Tom M Apostol. *Introduction to analytic number theory*. Springer Science & Business Media, 2013.

[7] D. F. Aranha and C. P. L. Gouvêa. RELIC is an Efficient LIbrary for Cryptography. http://code.google.com/p/relic-toolkit/.

[8] Diego F Aranha, Luis J Dominguez Perez, Amine Mrabet, and Peter Schwabe. Software implementation. *Guide to Pairing-Based Cryptography*, 2017.

[9] Paulo SLM Barreto, Ben Lynn, and Michael Scott. On the selection of pairing-friendly groups. In *International Workshop on Selected Areas in Cryptography*, pages 17–25. Springer, 2003.

[10] Paulo SLM Barreto and Michael Naehrig. Pairing-friendly elliptic curves of prime order. In *International Workshop on Selected Areas in Cryptography*, pages 319–331. Springer, 2005.

[11] Rana Barua, Ratna Dutta, and Palash Sarkar. Extending joux's protocol to multi party key agreement. In *Indocrypt*, volume 2904, pages 205–217. Springer, 2003.

[12] Mihir Bellare. New proofs for nmac and hmac: Security without collision-resistance. In *Annual International Cryptology Conference*, pages 602–619. Springer, 2006.

[13] Mihir Bellare, Ran Canetti, and Hugo Krawczyk. Hmac: Keyed-hashing for message authentication. *Internet Request for Comment RFC*, 2104, 1997.

[14] Mihir Bellare, Daniele Micciancio, and Bogdan Warinschi. Foundations of group signatures: Formal definitions, simplified requirements, and a construction based on general assumptions. In *Eurocrypt*, volume 2656, pages 614–629. Springer, 2003.

[15] Mihir Bellare and Phillip Rogaway. Random oracles are practical: A paradigm for designing efficient protocols. In *Proceedings of the 1st ACM conference on Computer and communications security*, pages 62–73. ACM, 1993.

[16] Mihir Bellare and Phillip Rogaway. Optimal asymmetric encryption. In *Workshop on the Theory and Application of of Cryptographic Techniques*, pages 92–111. Springer, 1994.

[17] Josh Benaloh. Verifiable secret-ballot elections. *PhD thesis, Yale University, Department of Computer Science Department*, 1987.

[18] Daniel J Bernstein, Johannes Buchmann, and Erik Dahmen. *Post-quantum cryptography*. Springer Science & Business Media, 2009.

[19] John Bethencourt, Amit Sahai, and Brent Waters. Ciphertext-policy attribute-based encryption. In *Security and Privacy, 2007. SP'07. IEEE Symposium on*, pages 321–334. IEEE, 2007.

[20] Eli Biham and Adi Shamir. Differential fault analysis of secret key cryptosystems. *Advances in Cryptology CRYPTO'97*, pages 513–525, 1997.

[21] Ian Blake, Gadiel Seroussi, and Nigel Smart. *Elliptic curves in cryptography*, volume 265. Cambridge university press, 1999.

[22] George Robert Blakley. Safeguarding cryptographic keys. In *Proc. AFIPS 1979 National Computer Conference*, pages 313–317, 1979.

[23] Lenore Blum, Manuel Blum, and Mike Shub. A simple unpredictable pseudo-random number generator. *SIAM Journal on computing*, 15(2):364–383, 1986.

[24] Manuel Blum, Paul Feldman, and Silvio Micali. Non-interactive zero-knowledge and its applications. In *Proceedings of the twentieth annual ACM symposium on Theory of computing*, pages 103–112. ACM, 1988.

[25] D Boneh, G Durfee, and Y Frankel. An attack on rsa given a fraction of the private key bits, asiacrypt 98. *Lecture Notes in Computer Science, Springer-Verlag, Berlin and New York*, 1998.

[26] Dan Boneh. The decision diffie-hellman problem. *Algorithmic number theory*, pages 48–63, 1998.

[27] Dan Boneh and Glenn Durfee. Cryptanalysis of rsa with private key d less than n/sup 0.292. *IEEE transactions on Information Theory*, 46(4):1339–1349, 2000.

[28] Dan Boneh and Matt Franklin. Identity-based encryption from the weil pairing. In *Advances in Cryptology-CRYPTO 2001*, pages 213–229. Springer, 2001.

[29] Dan Boneh, Eu-Jin Goh, and Kobbi Nissim. Evaluating 2-dnf formulas on ciphertexts. In *TCC*, volume 3378, pages 325–341. Springer, 2005.

[30] Dan Boneh, Antoine Joux, and Phong Q Nguyen. Why textbook elgamal and rsa encryption are insecure. In *ASIACRYPT*, volume 1976, pages 30–43. Springer, 2000.

[31] Dan Boneh, Ben Lynn, and Hovav Shacham. Short signatures from the weil pairing. *Advances in Cryptology ASIACRYPT 2001*, pages 514–532, 2001.

[32] Dan Boneh and Ramarathnam Venkatesan. Breaking rsa may be easier than factoring (extended abstract).

[33] Dan Boneh and Ramarathnam Venkatesan. Breaking rsa may not be equivalent to factoring. *Advances in Cryptology EUROCRYPT'98*, pages 59–71, 1998.

[34] Zvika Brakerski and Vinod Vaikuntanathan. Fully homomorphic encryption from ring-lwe and security for key dependent messages. In *Annual cryptology conference*, pages 505–524. Springer, 2011.

[35] Emmanuel Bresson, Olivier Chevassut, and David Pointcheval. Provably authenticated group diffie-hellman key exchange-the dynamic case. In *Asiacrypt*, volume 2248, pages 290–309. Springer, 2001.

[36] Emmanuel Bresson, Olivier Chevassut, and David Pointcheval. Dynamic group diffie-hellman key exchange under standard assumptions. In *Advances in Cryptology EUROCRYPT 2002*, pages 321–336. Springer, 2002.

[37] Emmanuel Bresson, Olivier Chevassut, David Pointcheval, and Jean-Jacques Quisquater. Provably authenticated group diffie-hellman key exchange. In *Proceedings of the 8th ACM conference on Computer and Communications Security*, pages 255–264. ACM, 2001.

[38] Emmanuel Bresson and Mark Manulis. Securing group key exchange against strong corruptions. In *Proceedings of the 2008 ACM symposium on Information, computer and communications security*, pages 249–260. ACM, 2008.

[39] Friederike Brezing and Annegret Weng. Elliptic curves suitable for pairing based cryptography. *Designs, Codes and Cryptography*, 37(1):133–141, 2005.

[40] Daniel RL Brown. Breaking rsa may be as difficult as factoring. *Journal of Cryptology*, 29(1):220–241, 2016.

[41] Francesco Buccafurri, Gianluca Lax, Rajeev Anand Sahu, and Vishal Saraswat. Practical and secure integrated pke+ peks with keyword privacy. In *e-Business and Telecommunications (ICETE), 2015 12th International Joint Conference on*, volume 4, pages 448–453. IEEE, 2015.

[42] Sanjit Chatterjee, Darrel Hankerson, and Alfred Menezes. On the efficiency and security of pairing-based protocols in the type 1 and type 4 settings. *WAIFI*, 6087:114–134, 2010.

[43] David Chaum. Blind signatures for untraceable payments. In *Advances in cryptology*, pages 199–203. Springer, 1983.

[44] David Chaum and Eugène Van Heyst. Group signatures. In *Advances in Cryptology - EUROCRYPT'91*, pages 257–265. Springer, 1991.

[45] Jie Chen, Hoon Wei Lim, San Ling, Huaxiong Wang, and Hoeteck Wee. Shorter ibe and signatures via asymmetric pairings. *Pairing*, 12:122–140, 2012.

[46] Kyu Young Choi, Jung Yeon Hwang, and Dong Hoon Lee. Efficient id-based group key agreement with bilinear maps. In *International Workshop on Public Key Cryptography*, pages 130–144. Springer, 2004.

[47] Henri Cohen. *A course in computational algebraic number theory*, volume 138. Springer Science & Business Media, 2013.

[48] Henri Cohen, Gerhard Frey, Roberto Avanzi, Christophe Doche, Tanja Lange, Kim Nguyen, and Frederik Vercauteren. *Handbook of elliptic and hyperelliptic curve cryptography*. CRC press, 2005.

[49] Craig Costello, Tanja Lange, and Michael Naehrig. Faster pairing computations on curves with high-degree twists. In *International Workshop on Public Key Cryptography*, pages 224–242. Springer, 2010.

[50] Craig Costello, Patrick Longa, and Michael Naehrig. Efficient algorithms for supersingular isogeny diffie-hellman. In *Annual Cryptology Conference*, pages 572–601. Springer, 2016.

[51] Joan Daemen and Vincent Rijmen. Rijndael for AES. In *AES Candidate Conference*, pages 343–348, 2000.

[52] Manik Lal Das, Ashutosh Saxena, and Deepak B Phatak. Algorithms and approaches of proxy signature: A survey. *arXiv preprint cs/0612098*, 2006.

[53] Angelo De Caro and Vincenzo Iovino. jpbc: Java pairing based cryptography. In *Computers and communications (ISCC), 2011 IEEE Symposium on*, pages 850–855. IEEE, 2011.

[54] Whitfield Diffie and Martin Hellman. New directions in cryptography. *IEEE transactions on Information Theory*, 22(6):644–654, 1976.

[55] Ratna Dutta, Rana Barua, and Palash Sarkar. Pairing-based cryptographic protocols: A survey. *IACR Cryptology ePrint Archive*, 2004:64, 2004.

[56] Iwan Duursma and Hyang-Sook Lee. Tate pairing implementation for hyperelliptic curves $y^2 = x^p - x + d$. In *Asiacrypt*, volume 2894, pages 111–123. Springer, 2003.

[57] Nadia El Mrabet and Marc Joye. *Guide to Pairing-Based Cryptography*. CRC Press, 2017.

[58] Taher ElGamal. A public key cryptosystem and a signature scheme based on discrete logarithms. *IEEE transactions on information theory*, 31(4):469–472, 1985.

[59] Uriel Feige, Dror Lapidot, and Adi Shamir. Multiple noninteractive zero knowledge proofs under general assumptions. *SIAM Journal on Computing*, 29(1):1–28, 1999.

[60] David Freeman, Michael Scott, and Edlyn Teske. A taxonomy of pairing-friendly elliptic curves. *Journal of cryptology*, 23(2):224–280, 2010.

[61] Steven D Galbraith et al. Supersingular curves in cryptography. In *Asiacrypt*, volume 2248, pages 495–513. Springer, 2001.

[62] Steven D Galbraith, Kenneth G Paterson, and Nigel P Smart. Pairings for cryptographers. *Discrete Applied Mathematics*, 156(16):3113–3121, 2008.

[63] Zvi Galil, Stuart Haber, and Moti Yung. Symmetric public-key encryption. In *Conference on the Theory and Application of Cryptographic Techniques*, pages 128–137. Springer, 1985.

[64] Michael R Garey and David S Johnson. *Computers and intractability*, volume 29. wh freeman New York, 2002.

[65] Daniel Genkin, Adi Shamir, and Eran Tromer. Rsa key extraction via low-bandwidth acoustic cryptanalysis. In *International Cryptology Conference*, pages 444–461. Springer, 2014.

[66] Daniel Genkin, Adi Shamir, and Eran Tromer. Acoustic cryptanalysis. *Journal of Cryptology*, 30(2):392–443, 2017.

[67] Craig Gentry. *A fully homomorphic encryption scheme*. Stanford University, 2009.

[68] Craig Gentry. Toward basing fully homomorphic encryption on worst-case hardness. In *CRYPTO*, volume 6223, pages 116–137. Springer, 2010.

[69] Craig Gentry and Shai Halevi. Implementing gentry's fully-homomorphic encryption scheme. In *EUROCRYPT*, volume 6632, pages 129–148. Springer, 2011.

[70] O. Goldreich, S. Micali, and A. Wigderson. How to play any mental game. In *Proceedings of the Nineteenth Annual ACM Symposium on Theory of Computing*, STOC '87, pages 218–229, New York, NY, USA, 1987. ACM.

[71] Oded Goldreich. Zero-knowledge twenty years after its invention. *IACR Cryptology ePrint Archive*, 2002:186, 2002.

[72] Oded Goldreich, Silvio Micali, and Avi Wigderson. Proofs that yield nothing but their validity or all languages in np have zero-knowledge proof systems. *Journal of the ACM (JACM)*, 38(3):690–728, 1991.

[73] Oded Goldreich and Yair Oren. Definitions and properties of zero-knowledge proof systems. *Journal of Cryptology*, 7(1):1–32, 1994.

[74] Shafi Goldwasser, Silvio Micali, and Charles Rackoff. The knowledge complexity of interactive proof systems. *SIAM Journal on computing*, 18(1):186–208, 1989.

[75] M Choudary Gorantla, Colin Boyd, and Juan Manuel González Nieto. Modeling key compromise impersonation attacks on group key exchange protocols. In *International Workshop on Public Key Cryptography*, pages 105–123. Springer, 2009.

[76] Vipul Goyal, Abhishek Jain, Omkant Pandey, and Amit Sahai. Bounded ciphertext policy attribute based encryption. *Automata, languages and programming*, pages 579–591, 2008.

[77] Vipul Goyal, Omkant Pandey, Amit Sahai, and Brent Waters. Attribute-based encryption for fine-grained access control of encrypted data. In *Proceedings of the 13th ACM conference on Computer and communications security*, pages 89–98. Acm, 2006.

[78] Robert Granger, Florian Hess, Roger Oyono, Nicolas Thériault, and Frederik Vercauteren. Ate pairing on hyperelliptic curves. In *EUROCRYPT*, volume 4515, pages 430–447. Springer, 2007.

[79] Jens Groth. Short pairing-based non-interactive zero-knowledge arguments. In *Asiacrypt*, volume 6477, pages 321–340. Springer, 2010.

[80] Jens Groth, Rafail Ostrovsky, and Amit Sahai. Non-interactive zaps and new techniques for nizk. In *CRYPTO*, volume 4117, pages 97–111. Springer, 2006.

[81] Jens Groth, Rafail Ostrovsky, and Amit Sahai. Perfect non-interactive zero knowledge for np. In *Eurocrypt*, volume 4004, pages 339–358. Springer, 2006.

[82] Jens Groth and Amit Sahai. Efficient non-interactive proof systems for bilinear groups. *Advances in Cryptology–EUROCRYPT 2008*, pages 415–432, 2008.

[83] Javier Herranz, Fabien Laguillaumie, Benoît Libert, and Carla Ràfols. Short attribute-based signatures for threshold predicates. In *Cryptographers Track at the RSA Conference*, pages 51–67. Springer, 2012.

[84] Jeffrey Hoffstein, Jill Pipher, and Joseph H Silverman. Ntru: A ring-based public key cryptosystem. In *International Algorithmic Number Theory Symposium*, pages 267–288. Springer, 1998.

[85] Yuval Ishai and Anat Paskin. Evaluating branching programs on encrypted data. In *TCC*, volume 4392, pages 575–594. Springer, 2007.

[86] Mitsuru Ito, Akira Saito, and Takao Nishizeki. Secret sharing scheme realizing general access structure. *Electronics and Communications in Japan (Part III: Fundamental Electronic Science)*, 72(9):56–64, 1989.

[87] David Jao and Luca De Feo. Towards quantum-resistant cryptosystems from supersingular elliptic curve isogenies. *PQCrypto*, 7071:19–34, 2011.

[88] David Jao and Vladimir Soukharev. Isogeny-based quantum-resistant undeniable signatures. In *International Workshop on Post-Quantum Cryptography*, pages 160–179. Springer, 2014.

[89] Antoine Joux. A one round protocol for tripartite diffie–hellman. *Journal of cryptology*, 17(4):263–276, 2004.

[90] M Kasahara, K Ohgishi, and R Sakai. Cryptosystems based on pairing. In *The 2001 Symposium on Cryptography and Information Security*, 2001.

[91] Jonathan Katz and Yehuda Lindell. *Introduction to modern cryptography*. CRC press, 2014.

[92] Jonathan Katz and Ji Sun Shin. Modeling insider attacks on group key-exchange protocols. In *Proceedings of the 12th ACM conference on Computer and communications security*, pages 180–189. ACM, 2005.

[93] Jonathan Katz and Moti Yung. Scalable protocols for authenticated group key exchange. In *Crypto*, volume 3, pages 110–125. Springer, 2003.

[94] Dalia Khader. Attribute based group signatures. *IACR Cryptology ePrint Archive*, 2007:159, 2007.

[95] N Koblitz. elliptic curve cryptosystem, mathematics of computation, volume 48-1987. Technical report, PP-203-209, 1987.

[96] Neal Koblitz, Alfred Menezes, and Scott Vanstone. The state of elliptic curve cryptography. In *Towards a quarter-century of public key cryptography*, pages 103–123. Springer, 2000.

[97] Paul Kocher, Joshua Jaffe, and Benjamin Jun. Differential power analysis. In *Advances in cryptology CRYPTO 99*, pages 789–789. Springer, 1999.

[98] Paul Kocher, Joshua Jaffe, Benjamin Jun, and Pankaj Rohatgi. Introduction to differential power analysis. *Journal of Cryptographic Engineering*, 1(1):5–27, 2011.

[99] Paul C Kocher. Timing attacks on implementations of diffie-hellman, rsa, dss, and other systems. In *Annual International Cryptology Conference*, pages 104–113. Springer, 1996.

[100] Paul C Kocher. Timing attacks on implementations of diffie-hellman, rsa, dss, and other systems. In *Annual International Cryptology Conference*, pages 104–113. Springer, 1996.

[101] Eyal Kushilevitz and Rafail Ostrovsky. Replication is not needed: Single database, computationally-private information retrieval. In *Foundations of Computer Science, 1997. Proceedings., 38th Annual Symposium on*, pages 364–373. IEEE, 1997.

[102] Hendrik W Lenstra Jr. Factoring integers with elliptic curves. *Annals of mathematics*, pages 649–673, 1987.

[103] Allison Lewko, Tatsuaki Okamoto, Amit Sahai, Katsuyuki Takashima, and Brent Waters. Fully secure functional encryption: Attribute-based encryption and (hierarchical) inner product encryption. In *Annual International Conference on the Theory and Applications of Cryptographic Techniques*, pages 62–91. Springer, 2010.

[104] Jin Li and Kwangjo Kim. Attribute-based ring signatures. *IACR Cryptology EPrint Archive*, 2008:394, 2008.

[105] Adriana López-Alt, Eran Tromer, and Vinod Vaikuntanathan. On-the-fly mul-tiparty computation on the cloud via multikey fully homomorphic encryption. In *Proceedings of the forty-fourth annual ACM symposium on Theory of computing*, pages 1219–1234. ACM, 2012.

[106] Ben Lynn. Pbc library manual 0.5. 11, 2006.

[107] Ben Lynn. *On the implementation of pairing-based cryptosystems*. PhD thesis, Stanford University Stanford, California, 2007.

[108] Hemanta K Maji, Manoj Prabhakaran, and Mike Rosulek. Attribute-based signatures: Achieving attribute-privacy and collusion-resistance. *IACR Cryptology ePrint Archive*, 2008:328, 2008.

[109] Hemanta K Maji, Manoj Prabhakaran, and Mike Rosulek. Attribute-based signatures. In *Cryptographer's Track at the RSA Conference*, pages 376–392. Springer, 2011.

[110] Davender S Malik, John N Mordeson, and MK Sen. *Fundamentals of abstract algebra*. McGraw-Hill, 1997.

[111] Masahiro Mambo, Keisuke Usuda, and Eiji Okamoto. Proxy signatures: Delegation of the power to sign messages. *IEICE transactions on fundamentals of electronics, communications and computer sciences*, 79(9):1338–1354, 1996.

[112] Stefan Mangard, Elisabeth Oswald, and Thomas Popp. *Power analysis attacks: Revealing the secrets of smart cards*, volume 31. Springer Science & Business Media, 2008.

[113] Ueli M Maurer and Stefan Wolf. The relationship between breaking the diffie–hellman protocol and computing discrete logarithms. *SIAM Journal on Computing*, 28(5):1689–1721, 1999.

[114] Robert J McEliece. A public-key cryptosystem based on algebraic. *Coding Thv*, 4244:114–116, 1978.

[115] Alfred J Menezes, Tatsuaki Okamoto, and Scott A Vanstone. Reducing elliptic curve logarithms to logarithms in a finite field. *IEEE Transactions on information Theory*, 39(5):1639–1646, 1993.

[116] Alfred J Menezes, Tatsuaki Okamoto, and Scott A Vanstone. Reducing elliptic curve logarithms to logarithms in a finite field. *iEEE Transactions on information Theory*, 39(5):1639–1646, 1993.

[117] Alfred J Menezes, Paul C Van Oorschot, and Scott A Vanstone. *Handbook of applied cryptography*. CRC press, 1996.

[118] Ralph Charles Merkle, Ralph Charles, et al. Secrecy, authentication, and public key systems. 1979.

[119] Victor S Miller. Use of elliptic curves in cryptography. In *Conference on the Theory and Application of Cryptographic Techniques*, pages 417–426. Springer, 1985.

[120] Victor S. Miller. Short programs for functions on curves. *Unpublished manuscript*, 1986.

[121] David Naccache and Jacques Stern. A new public key cryptosystem based on higher residues. In *Proceedings of the 5th ACM conference on Computer and communications security*, pages 59–66. ACM, 1998.

[122] National Bureau of Standards. Data encryption standard (DES). *Federal Information Processing Standards Publication*, 46(3):1–22, 1977.

[123] National Institute of Standards and Technology. Announcing request for candidate algorithm nominations for the advanced encryption standard (AES). *Federal Information Processing Standards Publication*, 62(177):48051–48058, 1997.

[124] National Institute of Standards and Technology. Advanced encryption standard (AES). *Federal Information Processing Standards Publication*, 197(441):1–47, 2001.

[125] Naoki Ogura, Go Yamamoto, Tetsutaro Kobayashi, and Shigenori Uchiyama. An improvement of key generation algorithm for gentry's homomorphic encryption scheme. In *International Workshop on Security*, pages 70–83. Springer, 2010.

[126] Tatsuaki Okamoto and Katsuyuki Takashima. Homomorphic encryption and signatures from vector decomposition. In *International Conference on Pairing-Based Cryptography*, pages 57–74. Springer, 2008.

[127] Tatsuaki Okamoto and Katsuyuki Takashima. Efficient attribute-based signatures for non-monotone predicates in the standard model. In *International Workshop on Public Key Cryptography*, pages 35–52. Springer, 2011.

[128] Rafail Ostrovsky, Amit Sahai, and Brent Waters. Attribute-based encryption with non-monotonic access structures. In *Proceedings of the 14th ACM conference on Computer and communications security*, pages 195–203. ACM, 2007.

[129] Payal V Parmar, Shraddha B Padhar, Shafika N Patel, Niyatee I Bhatt, and Rutvij H Jhaveri. Survey of various homomorphic encryption algorithms and schemes. *International Journal of Computer Applications*, 91(8), 2014.

[130] Jacques Patarin. Hidden fields equations (hfe) and isomorphisms of polynomials (ip): Two new families of asymmetric algorithms. In *International Conference on the Theory and Applications of Cryptographic Techniques*, pages 33–48. Springer, 1996.

[131] Geovandro CCF Pereira, Marcos A Simplício, Michael Naehrig, and Paulo SLM Barreto. A family of implementation-friendly bn elliptic curves. *Journal of Systems and Software*, 84(8):1319–1326, 2011.

[132] David Pointcheval and Jacques Stern. Security arguments for digital signatures and blind signatures. *Journal of Cryptology*, 13(3):361–396, Jun 2000.

[133] J-J Quisquater and Chantal Couvreur. Fast decipherment algorithm for rsa public-key cryptosystem. *Electronics letters*, 18(21):905–907, 1982.

[134] Jean-Jacques Quisquater, Myriam Quisquater, Muriel Quisquater, Michaël Quisquater, Louis Guillou, Marie Annick Guillou, Gaïd Guillou, Anna Guillou, Gwenolé Guillou, and Soazig Guillou. How to explain zero-knowledge protocols to your children. In *Conference on the Theory and Application of Cryptology*, pages 628–631. Springer, 1989.

[135] Jean-Jacques Quisquater and David Samyde. Electromagnetic analysis (ema): Measures and counter-measures for smart cards. *Smart Card Programming and Security*, pages 200–210, 2001.

[136] MO Rabin. Digital signatures and public-key encryption as intractable as factorization. Technical report, MIT, Technical Report, MIT/LCS/TR-212, 1979.

[137] Somindu Ramanna, Sanjit Chatterjee, and Palash Sarkar. Variants of water's dual system primitives using asymmetric pairings. *Public Key Cryptography–PKC 2012*, pages 298–315, 2012.

[138] KC Reddy and Divya Nalla. Identity based authenticated group key agreement protocol. In *International Conference on Cryptology in India*, pages 215–233. Springer, 2002.

[139] Oded Regev. On lattices, learning with errors, random linear codes, and cryptography. *Journal of the ACM (JACM)*, 56(6):34, 2009.

[140] Ronald Rivest, Adi Shamir, and Yael Tauman. How to leak a secret. *Advances in Cryptology-ASIACRYPT 2001*, pages 552–565, 2001.

[141] Ronald L Rivest, Len Adleman, and Michael L Dertouzos. On data banks and privacy homomorphisms. *Foundations of secure computation*, 4(11):169–180, 1978.

[142] Ronald L Rivest, Adi Shamir, and Leonard Adleman. A method for obtaining digital signatures and public-key cryptosystems. *Communications of the ACM*, 21(2):120–126, 1978.

[143] Kenneth H Rosen. *Elementary number theory and its applications*. Addison-Wesley, 1993.

[144] Amit Sahai, Brent Waters, et al. Fuzzy identity-based encryption. In *Eurocrypt*, volume 3494, pages 457–473. Springer, 2005.

[145] Rajeev Anand Sahu and Sahadeo Padhye. Id-based signature schemes from bilinear pairing: A survey. *Frontiers of Electrical and Electronic Engineering in China*, 6(4):487–500, 2011.

[146] Yu Sasaki and Kazumaro Aoki. Finding preimages in full md5 faster than exhaustive search. In *EUROCRYPT*, volume 5479, pages 134–152. Springer, 2009.

[147] Peter Scholl and Nigel P Smart. Improved key generation for gentry's fully homomorphic encryption scheme. In *IMA International Conference on Cryptography and Coding*, pages 10–22. Springer, 2011.

[148] Michael Scott. Faster pairings using an elliptic curve with an efficient endomorphism. In *INDOCRYPT*, volume 3797, pages 258–269. Springer, 2005.

[149] Michael Scott. On the efficient implementation of pairing-based protocols. *Cryptography and Coding*, pages 296–308, 2011.

[150] Siamak F Shahandashti and Reihaneh Safavi-Naini. Threshold attribute-based signatures and their application to anonymous credential systems. In *International Conference on Cryptology in Africa*, pages 198–216. Springer, 2009.

[151] Adi Shamir. How to share a secret. *Communications of the ACM*, 22(11):612–613, 1979.

[152] Adi Shamir. Identity-based cryptosystems and signature schemes. In *Workshop on the Theory and Application of Cryptographic Techniques*, pages 47–53. Springer, 1984.

[153] Adi Shamir and Eran Tromer. Acoustic cryptanalysis. *presentation available from http://www. wisdom. weizmann. ac. il/ tromer*, 2004.

[154] Claude E Shannon. A mathematical theory of communication, part i, part ii. *Bell Syst. Tech. J.*, 27:623–656, 1948.

[155] Claude E Shannon. Communication theory of secrecy systems. *Bell Labs Technical Journal*, 28(4):656–715, 1949.

[156] Paul Sheridan. Hyperelliptic curve cryptography. 2003.

[157] You-hui SHI and Wei-sheng LI. A survey of blind signature studies [j]. *Computer Engineering & Science*, 7:028, 2005.

[158] Peter W Shor. Algorithms for quantum computation: Discrete logarithms and factoring. In *Foundations of Computer Science, 1994 Proceedings., 35th Annual Symposium on*, pages 124–134. Ieee, 1994.

[159] Joseph H Silverman. *The arithmetic of elliptic curves*, volume 106. Springer Science & Business Media, 2009.

[160] Nigel P Smart and Frederik Vercauteren. On computable isomorphisms in efficient asymmetric pairing-based systems. *Discrete Applied Mathematics*, 155(4):538–547, 2007.

[161] Marc Stevens. Single-block collision attack on md5. *IACR Cryptology ePrint Archive*, 2012:40, 2012.

[162] Douglas R Stinson. *Cryptography: theory and practice*. CRC press, 2005.

[163] Qiang Tang and Dongyao Ji. Verifiable attribute based encryption. *IJ Network Security*, 10(2):114–120, 2010.

[164] Namita Tiwari and Sahadeo Padhye. Analysis on the generalization of proxy signature. *Security and Communication Networks*, 6(5):549–566, 2013.

[165] Vinod Vaikuntanathan. Computing blindfolded: New developments in fully homomorphic encryption. In *Foundations of Computer Science (FOCS), 2011 IEEE 52nd Annual Symposium on*, pages 5–16. IEEE, 2011.

[166] Marten Van Dijk, Craig Gentry, Shai Halevi, and Vinod Vaikuntanathan. Fully homomorphic encryption over the integers. In *Annual International Conference on the Theory and Applications of Cryptographic Techniques*, pages 24–43. Springer, 2010.

[167] G Wang, Q Liu, and J Wu. Hierachical attibute-based encryption for fine-grained access control in cloud storage services, in proc. acm conf. *Computer and Communications Security (ACM CCS), Chicago, IL*, 2010.

[168] Lingling Wang, Guoyin Zhang, and Chunguang Ma. A survey of ring signature. *Frontiers of Electrical and Electronic Engineering in China*, 3(1):10–19, 2008.

[169] Xiaoyun Wang, Yiqun Lisa Yin, and Hongbo Yu. Finding collisions in the full sha-1. In *Crypto*, volume 3621, pages 17–36. Springer, 2005.

[170] Xiaoyun Wang and Hongbo Yu. How to break md5 and other hash functions. In *Eurocrypt*, volume 3494, pages 19–35. Springer, 2005.

[171] Michael J Wiener. Cryptanalysis of short rsa secret exponents. *IEEE Transactions on Information theory*, 36(3):553–558, 1990.

[172] Huixin Wu and Feng Wang. A survey of noninteractive zero knowledge proof system and its applications. *The Scientific World Journal*, 2014, 2014.

[173] Tsu-Yang Wu, Yuh-Min Tseng, and Tung-Tso Tsai. A revocable id-based authenticated group key exchange protocol with resistant to malicious participants. *Computer Networks*, 56(12):2994–3006, 2012.

[174] Tsu-Yang Wu, Yuh-Min Tseng, and Ching-Wen Yu. A secure id-based authenticated group key exchange protocol resistant to insider attacks. *J. Inf. Sci. Eng.*, 27(3):915–932, 2011.

[175] Tao Xie and Dengguo Feng. Construct md5 collisions using just a single block of message. *IACR Cryptology ePrint Archive*, 2010:643, 2010.

[176] Song Y Yan. *Primality testing and integer factorization in public-key cryptography*. Springer, 2009.

[177] Andrew C. Yao. Protocols for secure computations. In *Proceedings of the 23rd Annual Symposium on Foundations of Computer Science*, SFCS '82, pages 160–164, Washington, DC, USA, 1982. IEEE Computer Society.

[178] Andrew Chi-Chih Yao. How to generate and exchange secrets. In *Proceedings of the 27th Annual Symposium on Foundations of Computer Science*, SFCS '86, pages 162–167, Washington, DC, USA, 1986. IEEE Computer Society.

[179] Jianjie Zhao, Dawu Gu, and M Choudary Gorantla. Stronger security model of group key agreement. In *Proceedings of the 6th ACM Symposium on Information, Computer and Communications Security*, pages 435–440. ACM, 2011.

Index